Tectonics as a Process in Architecture

Tectonics as a Process in Architecture

Yonca Hurol
Cyprus International University

Copyright © 2025 by John Wiley & Sons, Inc. All rights reserved, including rights for text and data mining and training of artificial technologies or similar technologies.

Published by John Wiley & Sons, Inc., Hoboken, New Jersey.
Published simultaneously in Canada.

No part of this publication may be reproduced, stored in a retrieval system, or transmitted in any form or by any means, electronic, mechanical, photocopying, recording, scanning, or otherwise, except as permitted under Section 107 or 108 of the 1976 United States Copyright Act, without either the prior written permission of the Publisher, or authorization through payment of the appropriate per-copy fee to the Copyright Clearance Center, Inc., 222 Rosewood Drive, Danvers, MA 01923, (978) 750-8400, fax (978) 750-4470, or on the web at www.copyright.com. Requests to the Publisher for permission should be addressed to the Permissions Department, John Wiley & Sons, Inc., 111 River Street, Hoboken, NJ 07030, (201) 748-6011, fax (201) 748-6008, or online at http://www.wiley.com/go/permission.

The manufacturer's authorized representative according to the EU General Product Safety Regulation is Wiley-VCH GmbH, Boschstr. 12, 69469 Weinheim, Germany, e-mail: Product_Safety@wiley.com.

Trademarks: Wiley and the Wiley logo are trademarks or registered trademarks of John Wiley & Sons, Inc. and/or its affiliates in the United States and other countries and may not be used without written permission. All other trademarks are the property of their respective owners. John Wiley & Sons, Inc. is not associated with any product or vendor mentioned in this book.

Limit of Liability/Disclaimer of Warranty: While the publisher and author have used their best efforts in preparing this book, they make no representations or warranties with respect to the accuracy or completeness of the contents of this book and specifically disclaim any implied warranties of merchantability or fitness for a particular purpose. No warranty may be created or extended by sales representatives or written sales materials. The advice and strategies contained herein may not be suitable for your situation. You should consult with a professional where appropriate. Further, readers should be aware that websites listed in this work may have changed or disappeared between when this work was written and when it is read. Neither the publisher nor authors shall be liable for any loss of profit or any other commercial damages, including but not limited to special, incidental, consequential, or other damages.

For general information on our other products and services or for technical support, please contact our Customer Care Department within the United States at (800) 762-2974, outside the United States at (317) 572-3993 or fax (317) 572-4002.

Wiley also publishes its books in a variety of electronic formats. Some content that appears in print may not be available in electronic formats. For more information about Wiley products, visit our web site at www.wiley.com.

Library of Congress Cataloging-in-Publication Data applied for:

Paperback ISBN: 9781394329229

Cover Design: Wiley
Cover Image: Courtesy of Yonca Hurol

Set in 9.5/12.5pt STIXTwoText by Straive, Chennai, India

SKY10110408_060325

To my brother

Contents

Preface *xv*
Acknowledgments *xix*

1 **Introduction** *1*
Literature Review *4*
  On Tectonics and Tectonics Theories *4*
  On Tectonics and Tectonic Theories *10*
  On Tectonic Affects *11*
  On Innovation in Architecture *13*
Research Problem *16*
Aim and Claim of this Book *19*
Methodology *20*
Description of Data *24*
Process of Data Analysis *25*
Drawing Process Maps for Tectonic Affects *28*
The Ways of Avoiding Bias *31*
Contributions of this Book *33*
Readership *34*

Part I **Context of the Building, Its Project, and Tectonic Characteristics** *37*

2 **Context of the Project and Building and the Early Story of the House** *39*
About the Political Economy of Property Ownership in a Conflict Zone *39*
Context of Building Design and Construction *50*
Quality of Construction in North Cyprus *54*

viii | *Contents*

Horror Stories about Building a House   *60*
Disasters, Climate, and Building Culture in North Cyprus   *63*
Physical and Social Environment of the Building   *65*
About the Site   *67*
Conclusion   *69*

3   Preliminary Design of the House and Tectonic Affects due to Changes in the Application Project and the Tendering Process   *73*
Preliminary Design Ideas and Main Tectonic Decisions   *74*
  A Living Area Opening in Four Directions   *77*
    Comfort   *78*
    Use of Leaf Plaster   *80*
    Accessibility   *80*
    Small and Economic   *81*
    Hidden at the Back   *82*
    Relationship with the Sea View   *82*
    Security   *83*
    Colors   *84*
  Being Open to the Contributions of Builders, Foremen, and Workers   *85*
  The Inclusion of Tectonic Details   *85*
  Some Subconscious Dimensions in the Design of the Monarga House   *87*
    Towards Minor Architecture   *87*
    Not Being Domestic   *88*
    Memories – The Invisible   *90*
Tectonic Characteristics of the Completed Building   *91*
Tectonic Affects and Innovative Attitudes due to Changes During the Application Project   *96*
  Changes During the Application Project   *97*
  Tectonic Affects due to Changes in the Application Project   *100*
  Innovative Attitudes in the Application Project Phase   *104*
Tectonic Affects and Innovative Attitudes due to Changes During the Tendering Process   *105*
  Tectonic Affects due to the Changes in the Tendering Process   *106*
  Innovative Attitudes During the Tendering Process   *107*
Conclusion   *107*

Contents | ix

Part II   Changes in the Construction Process of the Building and Tectonic Affects/Innovative Attitudes   *111*

4   Tectonic Affects and Innovative Attitudes due to Changes During Construction of the Foundations, Frame System, Walls, and Roof   *113*
Foundations – Changes, Tectonic Affects, and Innovative
    Attitudes   *113*
  Tectonic Affects due to Changes During the Construction of the
      Foundations   *116*
  Innovative Attitudes During the Construction of the
      Foundations   *121*
Frame System – Changes, Tectonic Affects, and Innovative
    Attitudes   *122*
  Changes During the Construction of the Reinforced Concrete
      Frame System   *123*
  Tectonic Affects due to Changes During the Construction of the
      Reinforced Concrete Frame System   *124*
Walls, Heat Insulation, Openings, Changes, Tectonic Affects, and
    Innovative Attitudes   *127*
  Changes while Building the Brick Walls and Making the
      Openings in Them   *128*
  Tectonic Affects due to Changes During the Construction
      of Walls, Heat Insulation, and Openings   *130*
  Innovative Attitudes During the Construction of Brick Walls,
      Openings, and Heat Insulation   *132*
Roof, Heat Insulation on the Ceilings and Pergolas – Changes, Tectonic
    Affects, and Innovative Attitudes   *132*
  Changes During the Construction of the Roof, Heat Insulation,
      and Pergolas   *134*
  Tectonic Affects due to Changes During the Construction of the
      Roof   *138*
Conclusion   *145*

5   Tectonic Affects and Innovative Attitudes due to Changes Relating to Plastering, the Electrical, and Mechanical Systems, the Ceramics, and Water Isolation   *149*
Plastering, Painting – Changes, Tectonic Affects, and Innovative
    Attitudes   *150*
  Changes During Plastering and Painting   *152*

x | Contents

Tectonic Affects due to the Changes During the Application of Plastering/Painting *152*

Innovative Attitude During Plastering and Painting *152*

Electrical System – Changes, Tectonic Affects, and Innovative Attitudes *153*

Changes During the Installation of the Electrical System *156*

Tectonic Affects due to Changes During the Placement of the Electric System *158*

Innovative Attitudes During the Application of the Electrical System *158*

Mechanical Systems – Changes, Tectonic Affects, and Innovative Attitudes *159*

Changes Made During the Placement of Mechanical System Elements *161*

Tectonic Affects due to Changes During the Placement of the Mechanical System *169*

Innovative Attitudes During the Application of Mechanical Systems *169*

Ceramics – Changes, Tectonic Affects, and Innovative Attitudes *171*

Changes During the Placement of Ceramics *173*

Tectonic Affects due to Changes During the Placement of Ceramics *175*

Innovative Attitude During the Application of Ceramics *177*

Water Isolation – Changes, Tectonic Affects, and Innovative Attitudes *178*

Changes During the Application of Water Isolation *178*

Tectonic Affects Caused by Changes During the Application of Water Isolation *180*

Innovative Attitudes During the Application of Water Isolation *180*

Conclusion *180*

6 Tectonic Affects and Innovative Attitudes due to Construction Changes Relating to Windows, Doors, Wardrobes and Cupboards, the Fireplace and Chimney *185*

Windows, Shutters, the Front Door – Changes, Tectonic Affects, and Innovative Attitudes *185*

Changes During the Placement of the Aluminum Windows, Shutters, and Front Door *186*

Tectonic Affects Caused as A Result of Changes in the Windows, Shutters, and Front Door *191*

Innovative Attitudes During Placement of the Windows, Shutters, and Front Door   *191*

Interior Doors – Changes, Tectonic Affects, and Innovative Attitudes   *195*

Changes During the Placement of Interior Doors   *195*

Tectonic Affects Caused due to the Changes in the Processes of Interior Doors   *198*

Innovative Attitudes During the Placement of Interior Doors   *202*

Cupboards and Wardrobes – Changes, Tectonic Affects, and Innovative Attitudes   *203*

Changes During the Positioning of Wardrobes and Cupboards   *204*

Tectonic Affects Caused due to Changes in the Production and Placement of Wardrobes and Cupboards   *206*

Innovative Attitudes During the Placement of Wardrobes and Cupboards   *209*

The Fireplace and Chimney – Changes, Tectonic Affects, and Innovative Attitudes   *209*

Changes During the Construction of the Fireplace and Chimney   *211*

Tectonic Affects Caused by Changes in the Production and Placement of the Fireplace and Chimney   *214*

Innovative Attitudes During Construction of the Fireplace and Chimney   *216*

Conclusion   *216*

7   **Tectonic Affects and Innovative Attitudes due to Changes During the Construction Work in the Garden**   *221*

Garden Walls – Changes, Tectonic Affects, and Innovative Attitudes   *222*

Changes During the Construction of Garden Walls   *223*

Tectonic Affects due to Changes During the Construction of Garden Walls   *225*

Innovative Attitudes During the Construction of Garden Walls   *227*

Drainage – Changes, Tectonic Affects, and Innovative Attitudes   *228*

Changes During the Organization of the Drainage   *230*

Tectonic Affects due to Changes During the Installation of Drainage   *231*

Innovative Attitudes During the Installation of Drainage   *232*

xii *Contents*

The Garage and Concrete Work in the Garden – Changes, Tectonic
    Affects, and Innovative Attitudes *232*
  Changes During the Construction of the Plinth Protection *234*
  Changes During the Construction of the Garage and Work with
    Concrete to Provide Pedestrian Footpaths *235*
  Tectonic Affects due to Changes During the Construction of the
    Garage and Concrete Work in the Garden *237*
  Innovative Attitudes During the Construction of the Garage
    and the Concrete Work in the Garden *237*
Fences, Dog Kennels, Garden Gates – Changes, Tectonic Affects,
    and Innovative Attitudes *239*
  Changes due to Metalwork in the Garden *239*
  Tectonic Affects due to Changes During the Construction and
    Placement of Metalwork in the Garden *243*
  Innovative Attitudes During the Construction and Placement
    of the Metalwork *243*
Pathways and Roads – Changes, Tectonic Affects, and Innovative
    Attitudes *243*
  Changes in the Pathways and Roads *243*
  Tectonic Affects due to Changes During the Construction
    of Roads and Pathways in the Garden *246*
  Innovative Attitudes During the Construction of the Roads and
    Pathways *248*
Timber Work in the Garden – Changes, Tectonic Affects, and
    Innovative Attitudes *248*
  Changes During the Execution of Timber Work in the
    Garden *249*
  Tectonic Affects due to Changes During the Construction
    of Timber Elements in the Garden *249*
  Innovative Attitudes During the Execution of the Timber Work
    in the Garden *250*
Conclusion *250*

**Part III Changes, Tectonic Affects, and Innovative Attitudes
within the Building Process *255***

**8 Process Maps for Tectonic Affects and Innovative Attitudes within the
Building Process *257***
The Map of Changes with Tectonic Affects for the Monarga
    House *257*
Impact of Changes on the Initial Design Ideas *266*

Changes with Innovative Attitudes *268*

Actors of Changes with Tectonic Affects During the Building
Process *270*

Theoretical Reflections for the Changes, their Tectonic Affects, and
Related Innovative Attitudes *272*

The Results of the Seven Interviews *273*

Conclusion *277*

9 Conclusion *279*

Objective Results of this Research *280*

Potential Biases in This Research and How to Avoid Them *283*

Alternative Theory Fragments About Procedural Tectonics *284*

On a Procedural Theory of Tectonics *286*

A Critique of Some Tectonic Theories *286*

Actors of Procedural Tectonics: Recognizing the Role of All Actors
in Building Processes *289*

Ethics and Building Production *292*

Reasons for Change in Building Processes *294*

Beneficial and Harmful Tectonic Processes: A Materialistic
Approach to Architecture *295*

Making Tectonic Processes Beneficial *295*

Instrumentalized Tectonic Processes *300*

Details as Loci of Tectonic Processes *304*

A Deeper Inquiry into the Procedural Approach to Theory of
Tectonics *313*

Preface

The research presented in this book has been, and continues to be, an important part of my private life. Because of this, it extended beyond my dedication to academia. I had the opportunity to live approximately nine years of my life in such a way that my desire to design and build a house (the Monarga House) for myself aligned perfectly with my academic research interest in tectonics in architecture.

At the beginning, taking notes on everything and photographing every little step felt like a game to me. At the end of each working day, I would spend half an hour writing down what I remembered and classifying and recording the photos taken that day. I enjoyed these activities. My initial intention was simply to have a diary of the house. However, recording everything also began to help me manage the complex process, which involved many people and products.

It didn't take long for me to realize that the content of my writings and photos overlapped with my 35 years of research. At that point, I began to consider writing a book on tectonics in architecture, as I was aware that the role of technical issues had almost been diminished in contemporary tectonics theories. I thought it would be worthwhile to write a theoretical book on tectonics, in which aesthetics/meaning and building technology would not overshadow one another. However, I also doubted whether this research, centered around a small house, could make a significant contribution. Nevertheless, I continued to record the process in detail.

During my 40-year academic career, I also worked on academic research methods in architecture, ethics in academic research, and architectural ethics, in addition to tectonics in architecture. This enabled me to conceptualize the data collection methodology which corresponded with my activity recording. I knew that there were qualitative research methodologies that could be based on such data. I was also interested in the ethical dimensions for all roles involved in this activity, such as contractor, architect, controller, builder and others.

Since we had signed a contract with the contractor, I initially believed that changes should be minimized. This made me sensitive to changes within the process, so I began collecting detailed information about every change that occurred. It did not take long for me to realize that changes are unavoidable and can sometimes have a positive or a negative impact on the building's tectonics. I soon recognized the considerable influence of the contractor's team, including builders, subcontractors, and workers, on the building. Therefore, I continued to record activities in detail.

Since the appreciation of all professional contributions to architecture of buildings, such as the contributions of architects working in architectural offices, has become an ethical issue in the field for more than 20 years, I was curious about the contributions of all actors to the tectonics and design of my house. I was well aware that the contractor's team had a considerable impact on my house. Changes in all activities caught my attention, particularly due to my interest in Open Building theory, which supports later changes in architecture. I am still one of the editors of *Open House International*, a journal founded in 1976 to publicize academic research on the Open Building approach.

Although I was legally separated from the initial contractor of the house, I realized that the builders/workers who completed my house also contributed significantly. Writing facilitated this awareness, as I initially tended to overlook their contributions. When I began analyzing the written data (observation notes or the house diary) about the changes that occurred throughout the process, I was shocked by the number of changes and the extent of the contributions of the contractor's and builders' teams. This was not a simple task. It required the analysis and classification of all changes, distinguishing those with tectonic affects, identifying innovative attitudes, and pinpointing their respective actors.

Many of the contractors' contributions were positive, though there were also some notably negative changes. At that moment, I realized that my close monitoring of the process allowed me to perceive something that would have otherwise gone unnoticed. It was then that I decided to write this book, intending to demonstrate that tectonics is a process.

While writing the diary, I realized that I had multiple roles in the process of building my house. I was the researcher, the owner, the neighbor (as I was living very close to the construction site), the architect, and the controller. To emphasize how closely I inspected the process and to highlight the plurality of my roles, I refer to myself as RONAC (which stands for Researcher, Owner, Neighbor, Architect, Controller) in this book.

Completing the first layer of analysis revealed that the tectonics of my house was procedural. After that, I decided to dig deeper to uncover more details about the innovative attitudes and the actors behind the changes. These findings led me to believe that the contributions of the contractor's team and also the builders/foremen/technicians/carpenters to tectonics and architecture deserve recognition in theories of tectonics and architecture. This is because, in these theories, architecture is often presented as a complete and frozen object.

I conducted three interviews with professionals who had the opportunity to closely observe a building's construction. This research reinforced the findings of my study. Additionally, I conducted four more interviews with experienced contractors, controllers, and the head of Chamber of Architects in North Cyprus. I discovered that they prefer all changes on construction sites to be approved by qualified professionals. Nevertheless, I sought a way to acknowledge the contributions of contractors, as well as builders/foremen/technicians/carpenters. I proposed a collaborative approach that takes into account the concerns of professionals advocating for professional decision-making on every issue. However, I believe my primary contribution to the recognition of contributions of builders/foremen/technicians/carpenters lies in establishing a foundation for a theory of tectonics that emphasizes architectural details.

I can say that this book concludes with a theory of tectonic details in architecture, but it also highlights many problems within building processes, such as not fully utilizing the potential of the actors involved and the risk of instrumentalization.

Acknowledgments

Collecting data for this research took more than two years, and writing the book took another four years. As will be evident from the content, there are many actors involved in both the building process and the research process. Numerous individuals provided help during these stages.

Some of these individuals are represented by their professional roles, such as the architect of the application project, civil engineers, a mechanical engineer, an electrical engineer, contractors, builders, foremen, technicians, carpenters, workers, and the project controller. The architect of the application project was supported by several employees who also contributed to the Monarga House project. Foremen, technicians, and carpenters were also involved in the construction process.

Other contributors are identified by code names, such as RONAC (Researcher, Owner, Neighbor, Architect, Controller), OAC (Owner, Architect, Controller), ONACC (Owner, Neighbor, Architect, Controller, Contractor), OCCC (Owner, Civil Engineer, Controller, Contractor). These individuals had the opportunity to closely observe a building's entire process. Sharing their memories of their experiences provided significant support to the book's central claims. Family members such as my son provided help too.

Two contractors are mentioned in the book. One was responsible for constructing three-quarters of the Monarga House, while the other participated as an interviewee, offering valuable insights and perspectives for the book. Additionally, two controllers are highlighted: the main controller of the Monarga House, who is also an interviewee identified as OCCC, made significant contributions to both the house and the research. The second controller, an architect, also participated as an interviewee. The head of the Chamber of Architects of North Cyprus also contributed to this research as an interviewee.

The book also references neighbors, as well as animals, trees, and plants around the house. On a larger scale, macro-level actors such as immigrant workers in

Acknowledgments

North Cyprus, and the others involved in various construction activities, are also discussed.

To express my gratitude to everyone who played a positive role in the building process of the Monarga House and in the creation of this book, I have chosen to acknowledge them collectively. I am deeply thankful to all of them.

1

Introduction

Although architecture has been conceptualized as frozen music,[1] change exists constantly in all architectural processes. This includes the preparation of preliminary and application projects, the construction process and the phases of use.[2] The book claims that changes in the application project, tendering process, and construction process can significantly impact the tectonics of buildings, the tectonic affects they induce and even their architecture.

This is a serious problem in the building sector, as an architectural project can diverge greatly from the completed building. Changes may occur in the plan, details, and even in the selected materials and technology due to availability issues, problems in the project, ambiguity in the application project's handling of certain issues and errors in the project. Depending on the political-economic context of the building, these changes can also become drastic. A friend of the author shared "People get shocked when they see their completed houses" because the outcome differs significantly from the original project. The same friend, who is having a house built, also mentioned, "The contractor misunderstood the site plan and changed the position of our house on the site. This changed the whole site plan."[3]

Caricatures even depict this issue, where a building undergoes substantial changes after the architect meets with the owner and client.[4] According to one caricature, the building changes again after the working drawings of the application project are prepared and ultimately undergoes further changes during construction. The final building might have little resemblance to the initial design and the impact of this situation on the building's owner depends on various

---

1 Johann Wolfgang von Goethe, *Conversations of Goethe with Eckermann and Soret*. (Smith: Elder & Co., 1850).
2 Changes during the phases of use and maintenance phases are beyond the scope of this book.
3 Dr. Ceren Boğaç is the source of this information, and she has granted permission for its use.
4 Alia Alaryan et al., "Causes and Effect of Change Orders on Construction Projects in Kuwait" *Journal of Engineering Research and Applications* 4, no. 7, 2014: 1–8.

Tectonics as a Process in Architecture, First Edition. Yonca Hurol.
© 2025 John Wiley & Sons, Inc. Published 2025 by John Wiley & Sons, Inc.

1 Introduction

factors. Change management literature suggests that the most likely source of change within these processes is the client.[5]

Rework, which is a reactive type of change, usually causes negative results, delays, and increased costs but it is a specific category of change.[6] However, change can also be proactive and changes within the building process can lead to positive outcomes for clients and contractors. Some changes may enhance tectonics of a building. Various factors influence whether these changes have a positive or negative tectonic impact. This book advocates for viewing tectonics as a process to encourage beneficial changes.

The scientific literature on rework and changes during the construction process of buildings does not cover the tectonic and aesthetic dimensions of such changes. An example of this kind of literature is about *change management in construction*. This literature mainly focuses on the technical and economic dimensions of the change orders, the reasons for change, and how to manage changes.[7] However, the presence of this literature demonstrates that changes are occurring within building processes worldwide.

The dynamic capabilities approach to office management takes change in the environment seriously and it also emphasizes that an architectural or construction firm cannot complete a project without considering change.[8] This approach is also suggested to be used in conjunction with Building Information Management (BIM) and its emerging technologies, which represent the future of architectural and construction activities.[9] However, these approaches and methods do not typically address tectonics. There is a need for change to serve various purposes, including improving tectonic qualities and fostering innovation. Since existing tectonics theories fail to address the procedural aspects, there is a need for a new approach to tectonic theory.

Current approaches to office and project management do not acknowledge changes made by the building team on site (by the contractor, sub-contractor, builder, foremen, and worker) and instead expect the process to be managed solely

---

5 Alaryan "Causes and Effect of Change Orders."

6 Ramin Asadi et al., "Analyzing Underlying Factors of Rework in Generating Contractual Claims in Construction Projects" *Journal of Construction and Engineering Management* 149, no. 6, 2023. Adnan Enhassi et al., "Factors Contributing to Rework and their Impact on Construction Projects Performance" *International Journal of Sustainable Construction Engineering and Technology* 8, no. 1, 2017: 12–33.

7 Alaryan "Causes and Effect of Change Orders."; Qi Hao et al., Change Management in Construction Projects *International Conference on Information Technology in Construction CIB W78 2008* (Santiago, Chile, 2008).

8 David J. Teece, *Dynamic Capabilities and Strategic Management.* (Oxford University Press, 2009).

9 Şenyaşa, Gaye; Hurol, Yonca; Tolga Çelik, (2025) "Unveiling the Profound Mindset Shift in AEC Business: The Subtle Impact of BIM" Archnet-IJAR, Early access.

1 *Introduction* | **3**

by professionals. However, majority of tectonics (or architectonics)[10] theories also ignore changes that happen during the building process, including the preparation of the application project, the tendering process, and the construction process. The author of this book believes that this is because the creators of tectonics theories do not advocate for such significant changes, whether these changes result in negative outcomes or lead to positive outcomes.

Although most of the theories of architecture and tectonics see architecture as "frozen" as it was the case within Ancient Greek aesthetics, there are a few sources on "tectonic process"[11] and "tectonic transformation"[12] that address the issues of change. Many sources highlight the problem of building production processes, as architects and engineers have to choose building components off-the-shelf and this stops them from working with the building industry, builders and craftsmen. This divide between the design team and the building industry eliminates most potential changes, which should have been welcomed because it is expected to be more economical and faster. This approach also minimizes innovations and improvements in the building's tectonics.[13] However, being open to change and innovations is wiser[14] from an anthropological point of view if all parties in the building process agree. The most appropriate discipline to study tectonic affects due to changes, which can cause negative or positive changes in the building process, as well as innovations, is architecture and the most appropriate area for this research is tectonics in architecture.

The objective of this book is to provide an in-depth examination of the changes that occurred during the process of a single building: the Monarga House, which is shown in Figure 1.1. The study demonstrates that numerous changes took place during this process, many of which affected the building's tectonic character. Some of these changes involved innovative approaches and various actors, including contractors, builders, foremen, technicians, carpenters, and workers, as well as the owner and the controller, initiating these changes. This unique, in-depth examination of the processes of the Monarga House was possible because the roles of the

---

10 Tectonics, or architectonics, differs from the geological concept of tectonics, which refers to the layers of the Earth. The geological concept is primarily used by geologists, civil engineers, and earthquake engineers. In contrast, the architectural concept of tectonics refers to the aesthetic use of building technology, a concept which originates from Ancient Greece.
11 Anne Beim, *Tectonic Visions in Architecture*. (Royal Danish Academy of Fine Arts, School of Architecture, 1999).
12 Anne Marie Due Schmidt, "Tectonic Transformation: The Architect as an Agent of Change" In *Proceedings from the Annual Symposium of the Nordic Association for Architectural Research* Kunstakadeiets Arckitektskoles Forlag, 2006: 130–137. http://www.arkitekturforskning.net/.
13 Schmidt, "Tectonic Transformation."
14 Alain Badiou wrote that being open to change is ethical. Similarly, Gaston Bachelard supported the need for innovation and emphasized its importance for human beings. Alain Badiou, *Logics of Worlds*. Trans. A. Toscano. (Continuum, 2009). Gaston Bachelard, *The Psychoanalysis of Fire*, Trans. A. C. M. Ross. 2nd Edition. (Beacon Press, 1987[1938]).

Figure 1.1 A photo of the Monarga House in 2024.

researcher, owner, neighbor, architect, and controller (RONAC) were combined in one person, allowing for a comprehensive view of all changes, even those that were revised multiple times during the building process. The collection of such detailed data provided a fresh perspective on tectonics as a process. What makes the Monarga House a valuable case study is not the characteristics of the house itself, but the ability to observe and record every detail of its process.

Literature Review

This subject necessitates a comprehensive review of the literature on tectonics, including the holistic concept of tectonics, major tectonics theories, and theories on affects and tectonic affects, as well as changes in building processes and innovative attitudes. It is also important to include theories on architecture that emphasize the use of building technologies, along with relevant theories on construction.

On Tectonics and Tectonics Theories

Tectonics (or architectonics) is the artistic or aesthetic utilization of building technology and all other physical aspects of the environment in architecture. Building technology encompasses materials, structural systems, mechanical systems, electrical systems, information technology, and construction details. These issues can manifest unique tectonic characteristics in specific buildings. For example, Alvar Aalto used timber in a tectonic manner in Villa Mairea, incorporating timber columns and suspending the staircase with timber elements. Santiago

Calatrava's buildings often have tectonic structural systems. The Pompidou Center in Paris exhibits tectonic mechanical and electrical systems, as these systems were intentionally designed as colorful façade elements. New York's Times Square is defined by building façades covered with screens, which contribute to the square's tectonic quality, especially at night. Tadao Ando designed unique window details without visible mullions, particularly in the Church of Light. In contrast, the more ambiguous concept of "all other physical aspects of the environment" includes elements such as light, topography, climate, and culture. Buildings can be designed in ways that allow light, topography, and climate to influence the tectonics of the building. Tadao Ando again exemplifies the use of light as a tectonic element in the design of the cross-shaped and hidden windows of the Church of Light. Frank Lloyd Wright's Falling Water is perhaps the best example of the tectonic use of topography, as it carefully integrates the building with the waterfall and surrounding rocks.

Many examples of traditional architecture worldwide approach climatic issues in a tectonic manner, designing building forms to achieve climatic comfort (for example, preventing sun light from entering during summer while allowing it in during winter). Renzo Piano's Jean-Marie Tjibau Cultural Center reinterprets the form and surface characteristics of traditional huts of the local people in a tectonic way, creating a mixture of cultural and tectonic qualities by using traditional and modern materials and forms together. It can also be argued that historical, traditional, and vernacular buildings from various cultures and at different locations possess distinct tectonic features. Even within a single country, traditional and vernacular buildings can exhibit very different tectonic characteristics. For example, the stone vernacular architecture of Mardin, the timber framed masonry houses of Safranbolu, and the adobe Harran houses are all distinctly different examples from Türkiye. These differences arise from the use of locally available materials, varying climates, and differing functional and social needs. Regardless of the focus of tectonics, there is always an aesthetic dimension involved in the application of these technical and physical aspects of architecture and these aesthetic qualities are typically harmonious with nature. This implies that the concept of tectonics encompasses a broad spectrum within architecture, emphasizing physical, practical, and poetic and aesthetic aspects simultaneously. Tectonics approaches architecture in a holistic and ontological manner.[15] It should also be noted that many examples of architecture employ multiple physical elements to achieve tectonic characteristics.

---

15 Yonca Hurol, *The Tectonics of Structural Systems – An Architectural Approach*. (Routledge, 2016).

1 Introduction

There are five major tectonic theories that develop concepts to evaluate and understand the tectonic qualities of buildings and architecture. These concepts and the theoreticians who developed these theories are as follows:

- Architect Karl Botticher in the nineteenth century introduced the concepts of *kernform* and *kunstrform*.[16]
- Architect Gottfried Semper, also in the nineteenth century, used the concepts of *dressing*, *knot*, *tectonic*, and *stereotomic*.[17]
- In the twentieth century, the architect Eduard Sekler employed the concepts of *tectonic* and *atectonic*.[18]
- Architect Kenneth Frampton, during the late twentieth century, introduced the concepts of *tectonic form*, *authenticity*, *legibility* and "being *ontological* or *scenographic*."[19]
- During the late twentieth century, Gevork Hartoonian provided explanations about the *ontological approach to time* and *montage*.[20]

Botticher examined Gothic cathedrals and Ancient Greek temples as a foundation for his theory of tectonics. His theories aimed to understand the relationship between structure, form, and ornamentation. His concept of *kernform* represents a visible structure. He exemplified this type of structure with Gothic cathedrals. The concept of *kunstform* expresses the aesthetic qualities of this visible structure. He argued that the structural and ornamental elements should be integrated, advocating that while structure is essential, ornament is not. He emphasized reason and logic in architecture, suggesting that the relationship between structure and form should be transparent and legible. It can be said that Botticher defined a building with a visible and aesthetically valuable structure. He influenced many later theorists with his ideas on the relationship between structure, form, and ornament.

---

16 Karl Gottlieb Wilhelm Botticher, "The Principles of the Hellenic and Germanic Ways of Building with Regard to their Application to our Present Way of Building." In *What Style Should We Build?* Eds: J. Bloomfield, K. Forster and T. Reese. (The Getty Center Publication Programs, 1992[1828]): 147–168.

17 Gottfried Semper, *The Four Elements of Architecture and Other Writings*, Trans. Harry F. Mallgrave and Wolfgang Herrmann. 2nd Edition. (Cambridge University Press, 2010[1851]). Gottfried Semper, *Style in the Technical and Tectonic Arts; or, Practical Aesthetics*, Trans. Harry F. Mallgrave. (Getty Research Institute, 2004[1860]).

18 Eduard Sekler, "Structure, Construction, Tectonics." In *Structure in Art and Science*, Ed: G. Kepes. (Wordpress, 1965): 89–95. accessed August 18, 2018. https://610f13.files.wordpress .com/2013/10/sekler_structure-construction-tectonics.pdf.

19 Brian Kenneth Frampton, "Rappel a l'Ordre: The Case for the Tectonic." In *Labour, Work and Architecture*, Ed: K. Frampton. (Phaidon Press, 2002): 91–103. Brian Kenneth Frampton, *Studies in Tectonic Culture – The Poetics of Construction in Nineteenth and Twentieth Century Architecture*. (MIT Press, 1995).

20 Gevork Hartoonian, *Ontology of Construction – On Nihilism of Technology and Theories of Modern Architecture*. (USA: Cambridge University Press, 1994).

Semper is one of the most influential theorists in the field of tectonics, known for his understanding which was both complex and open to the realities of architecture, yet simultaneously respectful and sound. According to him, tectonics is the expression of cultural identity, historical continuity, and aesthetic values through the careful and meaningful use of materials and techniques. Gottfried Semper identified four fundamental elements of architecture: the *mound* (earthwork, including excavation, foundations, and platform), the *hearth* (fireplace and gathering area), the *roof and framework* (light framed tectonic structures) and the *enclosure* (woven wall – the infill between frame elements).[21] This perspective shows that Semper considered the industries and crafts in relation to architecture. He illustrated these four elements with a detailed drawing of a simple Caribbean hut, strongly indicating that forms and techniques are deeply rooted in nature, climate, and the traditions and practices of the societies that create them.

For Semper, craftsmanship is crucial to tectonic expression. His concept of *dressing* involves building forms determined by surface treatment, which often carries cultural and symbolic meanings beyond the structure of buildings. He argued that dressing is integral to architectural expression and should naturally arise from the construction process and materials. Semper also advocated for a polychromatic approach, as seen in Ancient Greek architecture. He emphasized the importance of tents, soft surfaces in buildings, and the *knots* (joints) within these soft surfaces, suggesting that humans can develop soft, harmonious relationships with soft surfaces.[22] Semper's *tectonic* and *stereotomic* concepts express the differentiation between light and heavy structures and buildings. A light structure can be achieved using timber, iron, or steel, while a heavy structure might be stone, brick, adobe masonry, or reinforced concrete. Semper proposed tectonic architecture, characterized by light and elemental structures, through his drawing of the Caribbean House, considering human emotions and the relationship between nature, culture, and architecture. While he supported advancements in building technology, he also critiqued the challenges that new architecture posed to nature and cultural traditions.

Sekler was influenced by both Botticher and Semper. According to him, tectonics is about expression through the structural system, construction, methods and materials. He believed that the structural system and construction methods should be visible, meaningful, aesthetic and clearly understandable to the observer. Making the logic of structural and construction legible and integral to aesthetic expression was particularly important to him. He argued that technical honesty, cultural expression, and architectural meaning should be combined.

---

21 Frampton, *Studies in Tectonic Culture*: 85.
22 Semper, *The Four Elements of Architecture*.

1 Introduction

Sekler developed the concept of *atectonic* to describe the illusory use of building technology, where structure and materials are concealed, surface appearance and decorative elements are prioritized, materials are used contrary to their qualities, form is disconnected from structural reality and/or construction elements are masked. If architectural form negates the expression of construction logic, it is also considered *atectonic*. Any disconnection between how a building looks and how it is built is undesirable. For this reason, if a building appeared structurally unbalanced, Sekler referred to it as *atectonic*. He opposed concealing the realities of technology and advocated for a tectonic use of building technology, where the components of technology are visible and understandable.

Kenneth Frampton criticizes decontextualized, industrialized, and commercialized aspects of modern architecture. His theories on tectonics parallel his broader architectural theory, which emphasizes an existential approach that values the environment, culture, and history. An *ontological approach to architecture* seeks to explore the fundamental nature of architecture by achieving existential qualities such as materiality, connection to the earth, and a significant role in human experience. The way buildings are grounded in their context is crucial, requiring openness to the deeper existential realities of human life. Tectonics is integral to this *ontological approach*. Frampton defines tectonics as the art of construction, as well as the poetics of construction. He emphasizes that *tectonic form* emerges from structural and material logic, where the structural framework should be legible and understandable. The way a building is construed should be apparent, with structure and construction serving as sources of architectural meaning. Authenticity, which involves a deep concern for materiality, is the genuine expression of materials, construction methods, and their relationship with context. The construction process should inform a building's form and aesthetics, clearly exhibiting the logic of its construction. An *ontological* structure possesses *autonomy* and adheres to *tectonic form*. In contrast, *scenographic* architecture aims to create visual impact, often producing striking spectacles (even through illusions) at the expense of structural integrity and material authenticity. This approach prioritizes representation and conceals the logic of construction.

Hartoonian also adopts an ontological approach to architecture, with his tectonic theories primarily influenced by Semper. He argues that a more existential approach is necessary to address the factors shaping contemporary architecture, such as political issues, economic concerns, and the influences of global capital. According to Hartoonian, architecture is not merely a visual object, it must be considered in its entirety. Tectonics derives its strength from the poetic potential of construction, encompassing not only technical and aesthetic issues but also the historical reflection of cultural values and the ideological underpinnings of a particular period. It is not just about structural expression but also about creating spaces that resonate with human experience on multiple levels, including sensory, emotional, and cultural. Tectonics represents the intersection of material reality and cultural imagination in architecture, requiring the articulation of structure and materials to convey cultural and historical meanings. A dialogue between form and structure is essential, allowing form to emerge from structural logic while considering cultural and historical context. The relationship between form and structure should be transparent and expressive, making the construction process legible and meaningful. The visible expression of construction elements connects the physical act of building with cultural narratives grounded in truthfulness. Ornament can contribute to tectonics by mediating between structure, construction logic, cultural meaning, and context, provided it is rooted in tectonic logic.

Hartoonian's concept of *ontological approach to time* does not reject the influence of historical architecture on new buildings. Instead, he advocates for the reinterpretation of historical issues in architecture, addressing aspects such as mass organization, the architectural role of structural elements, the redevelopment of historical building techniques, the reinterpretation of historical building components and the use of building technology to fulfill the values behind different functions. He neither suggests turning back to history nor solely focusing on the future. Hartoonian extends Semper's concepts of *joint-disjoint* and *montage*, applying them to various architectural cases, giving equal importance to both to technical and technological and artistic and aesthetic dimensions.[23] An example for montage (joint-disjoint) is Adolf Loos' Looshouse, where the upper floors, designed to resemble masonry with fewer windows for privacy, contrast with the lower floors, which openly display the frame system to enhance the visibility of the commercial facilities inside. This juxtaposition of characteristics in the Looshouse exemplifies the logic of montage. Hartoonian distinguishes himself from other theoreticians of tectonics by embracing the coexistence of contradictory characteristics in architecture and by not prioritizing either the technological or the artistic dimensions over the other.

When examining these theoreticians, it becomes clear that the theories of tectonics encompass two main dimensions. The first is the ontological and *existential dimension*, which views tectonics within a broader context of architecture. The second is the *ethical dimension*, which emphasizes the understandability, visibility, and honest use of building technology, also including its relationship with ornamentation.

There are also some new theories of tectonics emerging in the field. Chad Schward's work on tectonics primarily examines the construction of specific buildings through a phenomenological approach, highlighting the role of tectonics

---

23 Hartoonian, *Ontology of Construction.*

1 Introduction

in architecture.[24] Similarly, Alexis Gregory's work considers construction as a key tectonic issue that contributes to the overall architectural experience.[25] The primary contribution of these new theoreticians lies in their focus on the meaning and aesthetics of construction.

On Tectonics and Tectonic Theories

Most of the tectonic theories discussed above were significantly shaped by architectural developments arising from the Arts and Crafts Movement, Constructivism, the Bauhaus Movement, and High-Tech architecture. These architectural movements critically examined the relationship between technical issues and the practice of architecture. Therefore, it is essential to consider these movements as well to fully understand the evolution and impact of tectonic theories.

The influence of the nineteenth-century Arts and Crafts Movement on most theories of tectonics is evident. The main principles of the Arts and Crafts Movement included an emphasis on craftsmanship, simplicity with an avoidance of unnecessary ornamentation, the use of natural materials, the integration of arts and crafts, and meticulous attention to detail, coupled with a rejection of industrialization. The movement also embraced social ideals and an ontological approach to context, advocating for a return to traditional building practices where design and construction are more closely integrated.[26] Additionally, the Arts and Crafts Movement emphasized the dignity of labor and the importance of meaningful work for society.

Early twentieth-century Russian Constructivism advocated for a collaborative approach from the initial design phase through the completion of construction. The movement defended collective creativity and integrated industrial building techniques, standardization, and prefabrication into their projects. They opposed ornamentation and many of their projects were highly experimental. The goal was to reflect the emerging social order in their country through their building activities.[27] Russian Constructivism significantly influenced the Bauhaus Movement in Germany.

The Bauhaus Movement of the twentieth century was influenced by both the Arts and Crafts Movement and Constructivism. The Bauhaus advocated for

---

24 Chat Schward, *Introducing Architectural Tectonics – Exploring the Intersection of Design and Construction.* (Routledge, 2016).

25 Alexis Gregory, *Comprehensive Tectonics – Technical Building Assemblies from Ground to the Sky.* (Routledge, 2019).

26 William Morris, *The Collected Works of William Morris.* (Cambridge University Press, 2013[1910]).

27 Sima Ingberman, *ABC: International Constructivist Architecture 1922–1939.* (MIT Press, 1994).

the unity of art, craft, and technology, emphasizing collaboration and collective work. They viewed the designer as a craftsman and promoted standardization and prefabrication, simplicity in design, and an emphasis on material honesty.[28]

High-tech architecture is a twentieth-century approach that emphasizes the structural system and construction elements as primary considerations for creating performative tectonics in building design. High-tech buildings typically feature unique and specialized structural systems, with meticulously designed details. Many of these buildings incorporate innovations that set them apart from ordinary constructions. For example, the process behind the Pompidou Center in Paris was unique, as Peter Rice of Ove Arup and Partners played a crucial role in designing its innovative structural system, significantly contributing to the architecture of Renzo Piano and Richard Rogers. The gerberettes, which connect the columns to the 2D trusses, were designed by Peter Rice, cast in factory and installed by specialized worker teams who had to employ innovative construction techniques.

On Tectonic Affects

The concept of *affect* plays a central role in the aesthetics of Deleuze and Guattari.[29] It can be simply defined as the feelings created by physical things or objects, often without conscious awareness. Affects originate from matter and practice. The philosopher Baruch Spinoza categorized affects into the feelings of pleasure and pain, which encompass all feelings. Deleuze and Guattari categorized affects into three: pleasure, pain, and desire. Spinoza believed that desire is a kind of pleasure. This book follows Spinoza's categorization.[30]

Affects are practical psycho-physiological constructs, setting them apart from other physical effects. *Tectonic affects* are those caused by the tectonic characteristics of architecture. The concept of affect can be studied as a part of aesthetics, alongside composition principles, poetics, symbolism, representation, and so on. However, the concept of affect differs from these concepts because affects are individual emotions that are inherent to life, objects, things, matter, and experiences. In contrast, symbolism, and representation possess a transcendental and social character. Affects also differ from abstract composition principles.[31]

---

28 Herbert Bayer, *Bauhaus 1919–1928*. (MOMA, 1938).
29 Gilles Deleuze and Felix Guattari, *A Thousand Plateaus, Capitalism and Schizophrenia*. 5th edition, Trans. Brian Massumi Edition. (Continuum, 2004[1980]). Gilles Deleuze and Felix Guattari, *Anti-Oedipus – Introduction to Schizoanalysis*, Trans. Eugene Holland. (Routledge, 2002[1972]).
30 Baruch Spinoza, *The Ethics*. (Hackett Publishing, 1992[1677]). Deleuze and Guattari, *A Thousand Plateaus*; Ulus Baker In *From Opinions to Images: Essays towards a Sociology of Affects*, Eds: A. Ozgun and A. Treske. (Institute of Network Cultures No. 37, 2020).
31 Simon O'Sullivan "The Aesthetics of Affect – Thinking Art beyond Representation" *Angelaki – The Journal of the Theoretical Humanities* 6, no. 3, 2001: 125–135.

1 Introduction

Affects can cause arousal or motivation which can sometimes be so strong that it prompts immediate action by a person. For example, if someone feels that they are in danger due to certain affects, they would leave that place as soon as possible.[32] Affects typically induce unconscious emotions due to inherent inclinations in human psychology, such as the inclination to freedom, sex, power, and so on.[33] Emotions originating from tectonic affects can be felt by anyone, but architects are generally more aware of tectonic affects and their influence on architectural design. The degree of consciousness among architects regarding tectonic affects largely depends on their intellectual engagement with tectonics theories.

The importance of affects, which provide a body-mind continuum, lies in shaping the opinions (or beliefs) of people that develop through the influence of affects. In the age of images, numerous small likes and dislikes (feelings resulting from affects) accumulate to form opinions in people's minds. These opinions are not thoughts or philosophical ideas because they are not based on reason. However, they have become more powerful than reason in the contemporary postmodern world. Architecture also plays a role in shaping opinions, alongside the influence of media and social media.[34]

Yonca Hurol has a theory of *tectonic affects* in architecture.[35] Since affects are grounded in practical and material issues that lead to psychological outcomes, she analyzed tectonic affects in architecture as poetic affects, affects related to change and time, and affects related to domination. These categories emphasize the role of affects in the subjective and ideological dimension of opinions. She classifies tectonic affects into the following categories:

- Poetic tectonic affects
  - Through materials and systems (lightness and heaviness, hapticity, tectonic details)
  - Through details about functionality (human scale, corners and places, hiding places in system details)
  - Through continuity within the context (considerations of natural, historical, urban contexts and the affect of "house and universe")
- Tectonic affects of change (technical improvements, innovations, affirmative approaches, new or familiar, practicality)

---

32 O'Sullivan "The Aesthetics of Affect": 125–135. Leila Scannel and Robert Gifford "Defining Place Attachment: A Tripartite Organizing Framework" *Journal of Environmental Psychology* 30, 2010: 1–10. Maria Flores, accessed March 24, 2019. https://mariafloresarch.com/Affect-in-Architecture; Stephen Mackie, Psychospace Master's Thesis, Dundee University, 2010, accessed March 24, 2019. https://issuu.com/stephenmackie/docs/psychospace/48.

33 Immanuel Kant, *Lectures on Anthropology*. (Cambridge University Press, 2012[1798]).

34 Baker, *From Opinions to Images*: 10–12.

35 Yonca Hurol, *Tectonic Affects in Contemporary Architecture*. (Cambridge Scholars Publishing, 2022).

- Tectonic affects of time (timeliness, ontological approach to time, rejection of history, conservatism, futurism)
- Tectonic affects of domination
  - At the urban scale (expression of power, territorialization, aestheticization of politics, culture industry, sensational image-making through tectonics)
  - At the building scale (domination of form, thing (close to people or object) (distant to people), precision or imprecision)

Hurol also argues that the preference for these tectonic affects during architectural design depends on the function and context of the building, as well as the subjective preferences of the owner, architect, and/or construction firm. She advocates a moody approach to tectonics and tectonic affects in architecture depending on the context.

Chapter 3 of this book applies the above theories and concepts of tectonics and architecture to the Monarga House. This analysis is necessary to understand how changes during the building process influenced its tectonic characteristics with respect to initial definitions in the preliminary architectural project.

On Innovation in Architecture

Innovations in building activities can be defined as contributing to knowledge about buildings (technology, design, management) by developing new items and knowledge about them. There are some sources in which innovations are categorized as innovations in building materials, building systems (e.g., façade systems), the use of IT systems in architecture and building process management.[36] However, there can be small innovations as well as some famous innovations in architecture. It is useful to categorize innovations in a more detailed way to make people feel that there can be many different types of innovations. The number of the examples in the following list can be increased:

- Innovations in structural materials – such as the innovation of Ultra High-Performance Fiber Reinforced Concrete (UHPFRC), carbon fiber, titanium
- Innovations in other building materials – such as the innovation of ETFE (Ethylene Tetrafluoroethylene), PTFE (Polytetrafluoroethlene)
- Innovations in building form – such as the innovative formlessness of Frank Gehry's Guggenheim Museum, parametric forms of designs of Zaha Hadid Architects
- Innovations in the activities of structural engineers – such as the innovations made by Peter Rice to complete the structural analysis of the Sydney Opera

---

36 Eduardo Gutierrez et al., *Innovation Architecture*. (Actar Publisher, 2021). Andrew H Dent and Leslie Sherr, *Material Innovation – Architecture*. (Thames & Hudson, 2014). Ajla Aksamija, *Integrating Innovation in Architecture*. (John Wiley & Sons, 2016). Simon Vamvakidis, *Innovative Architecture Strategies*. (BIS Publishers, 2017).

House, to contribute to design of high-tech examples such as Pompidou Center, and so on

- Innovation of new structural systems – such as the innovation of tubular structures in 1960s by structural engineer Fazlur Khan and SOM (Skidmore, Owings, & Merrill), and suspended glass systems in 1990s again by Peter Rice
- Innovative reconsideration of a certain structural system – such as the special design of the space frame structure of the Beijing Water Cube by using structural units in the form of soap bubbles
- Innovation of a supplementary structural system – such as the use of belt trusses and outrigger trusses in high-rise building structures to increase their height
- Innovation of a special structural element – such as the innovation of mega columns and exoskeleton columns …
- Re-innovation of a forgotten building technique – such as architect Hassan Fathy's re-innovation of an old adobe dome building technique through reading old Egyptian papyrus documents
- Innovation of structural details – such as Peter Rice's design of the details of suspended glass systems as well as the damping details in high-rise building structures
- Innovation of building components – such as sustainable façade systems, kinetic façade systems
- Innovation of building techniques – such as innovation of the climbing form-work systems
- Innovation of mechanical systems – such as new heating cooling systems
- Innovation of new electrical systems
- Innovation of new IT systems
- Innovation of new fire safety systems
- Innovation of new acoustic systems
- Innovation of new foundation systems
- Innovations about green architecture
- Innovations about building furniture … etc.

These innovations occur at various stages of the building process. For example, the innovation of ETFE material was essential to enable the Beijing Water Cube while considering fire safety measures. ETFE material melts in the event of a fire but does not burn. This innovation was realized during the application project of the building, which was prepared by a specialized firm known for solving building problems through innovations. Innovations in projects by Zaha Hadid Architects take place at various levels. Some innovations are realized by computer engineers, enabling architects to explore new forms, while others involve mechanical engineering innovations that allow for 3D printing of building parts and so on. Peter Rice developed structural analysis through physical modeling as an innovation during the structural engineering phase of the Sydney Opera House. He also innovated suspended glass systems in response to architect Bernard Tschumi's demand for minimal structural elements in the transparent façades of the Paris Science Museum. Structural engineer Fazlur Khan innovated tubular structures to economically achieve taller skyscrapers. These are some of the very famous innovations. However, there are many less prominent innovations in various buildings that contribute to the diversity of the built environment. Contractors, builders, or workers can exhibit innovative attitudes to solve problems during construction and their solutions can be accepted by all parties in the process. Controllers might also develop innovative attitudes towards problems during construction, which can be recognized by others. Architects can exhibit innovative attitudes during the preliminary design, application project, and throughout the construction of the building. The concept of "innovative attitude" is particularly relevant to the subject of this book, which explores the processes of a simple house. An innovative attitude can be defined as being open to change, where change may involve using different materials that are not necessarily new, suggesting different systems, proposing alternative details, and solving problems in unique ways.

Many of these innovations and the buildings in which they were applied, are outcomes of procedural approaches. The tectonic qualities of these buildings are also the result of such approaches. For instance, the Beijing Water Cube is a product of a procedural approach, where its structural system and façade material were specifically innovated for this building, with the construction site serving as a space for trial and error. Similarly, structural engineer Peter Rice's contributions to the design of Sydney Opera House and innovation of suspended glass systems required close collaboration with contractors, builders, and foremen. In projects by Zaha Hadid Architects, 3D printing of building parts is an outcome of collaborative efforts between computer specialists, architects, and engineers on site, where computer specialists work alongside builders, workers, and designers. Hassan Fathy's re-innovation of old shallow adobe domes was based on very simple principles, with workers actively contributing to both in the design and construction processes.

These cases are often viewed as special and radically different from ordinary buildings. However, the production of ordinary buildings also has a procedural character, which similarly affects their tectonics. However, this influence is often unacknowledged. Some architectural approaches, such as Edwards Ford's ontological approach to construction, present details as integral parts of both the project and the completed buildings. While detailed drawings reveal what is inside a detail, they cannot convey the procedural character of construction, particularly in terms of changes in materials, applications, and other variables.[37] Additionally, there are no architectural or tectonics theories that sufficiently support or address the procedural approach to tectonics.

Research Problem

This book claims that tectonics is influenced by the entire building process, as this approach was introduced by Anne Beim and Anne Marie Due Schmidt.[38] Gottfried Semper, Kenneth Frampton, and Gevork Hartoonian have written about the role of building processes in tectonics, but they have not explored how this occurs in detail. Current architectural and tectonic theories do not support a procedural approach to tectonics. However, all aspects of building processes (including their application projects, tendering, construction processes such as the construction of foundations, structural systems, walls, roof systems, and finishing details) can significantly impact the tectonics of buildings. Figure 1.2 illustrates the initial east façade of the Monarga House, together with various impacts caused by building processes (impact of application project, construction of foundations finishes, closures, mechanical and electrical systems, landscaping, interior design, etc.), which are represented as forces. These impacts are mainly caused by ambiguities within the project.

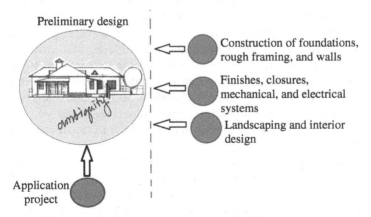

Figure 1.2 Process of tectonics covering the whole building process.

---

37 Edward Ford, *Five Houses, Ten Details*. (Chronicle Books, 2009).
38 Beim, *Tectonic Visions in Architecture*; Schmidt, "Tectonic Transformation."

Since the claim of this book is about the procedural nature of tectonics in architecture, it suggests that contractors, builders, foremen, technicians, and carpenters also make significant contributions to tectonics. Therefore, the research problem should be framed in relation to the issues that highlight the procedural character of tectonics.

Semper emphasized that consideration of the inherent qualities of materials and craftsmanship is crucial for achieving tectonic expression. He primarily referred to craftsmen who *dress* huts by weaving coverings for their surfaces.[39] Frampton focused on existential qualities, such as the role of buildings in human experience, their grounding in context, and their relationship to deeper existential aspects of human life, which he considers to be the base of ontological issues of tectonics. He defines tectonics as the expressive potential of materials, construction methods, and craftsmanship in conveying architectural meaning connected to culture and history.[40] Hartoonian argued that tectonics is not merely about structural expression but also about creating spaces that resonate with human experience on multiple levels, including the experience of builders, foremen, and workers. All three (Semper, Frampton, and Hartoonian) emphasize the role of craftsmanship, particularly in relation to the cultural and historical aspects of tectonics. Hartoonian specifically uses the term *metier* to refer to the interpretation of historical details and techniques in construction.[41]

These theoreticians presented the role of a craftsman as one who collaborates with professionals and contributes to the design process in architecture. However, they do not specifically address the roles of contractors, builders, foremen, technicians, and carpenters. While some of these theories touch on aspects of the building process (such as the contributions of craftsmen and the use of techniques like concealing elements or creating illusions in building) they primarily focus on architecture as a finished product or outcome. As a result, there is insufficient consideration of the procedural character of tectonics in architecture. This book argues that recognizing the procedural nature of tectonics could improve the tectonic qualities of buildings, foster innovation, and decrease instrumentalized approaches to building activities.

The Arts and Crafts Movement was more focused on the contributions of craftsmen than on the theories of tectonics, as it advocated a return to traditional building practices where design and construction were more integrated and craftsmanship was highly appreciated.[42] Architects were encouraged to work directly with craftsmen, who were seen as skilled artists. The Arts and Crafts

---

39 Semper, *The Four Elements of Architecture.*
40 Frampton, *Studies in Tectonic Culture.*
41 Hartoonian, *Ontology of Construction.*
42 William Morris, *The Collected Works of William Morris.* (Cambridge University Press, 2013[1910]).

18 | *1 Introduction*

Movement also emphasized the dignity of labor and the importance of meaningful work in society. This book does not advocate for a return to tradition. Instead, its intention is to acknowledge the procedural character of tectonics in modern building practice and to develop a foundation within tectonics theory to make this more beneficial.

Early twentieth-century Constructivism also considered architecture as a collaborative medium, in which professionals, builders, and craftsmen worked together from the initial design phase through the end of construction. Both professionals and builders, foremen, workers, and craftsmen were expected to be equipped with both knowledge and practical skills. The movement incorporated industrial building techniques, standardization, and prefabrication.[43] This book suggests recognizing the contributions made during the building process by the various actors, with a focus on their knowledge and practical skills.

Unlike Constructivism, the Bauhaus Movement emphasized the role of the designer as a craftsman within modern building practices and suggested collaborative work between them. In contrast to the Arts and Crafts Movement, the Bauhaus placed a greater emphasis on professionals, contemporary science, techniques, and technologies.[44] These movements reflect the evolving roles of those involved in the production of buildings (including professionals, contractors, builders, foremen, technicians, and craftsmen) within the recent history of architecture.

The recent studies on tectonics in construction, such as the works of Chad Schwartz,[45] Alexis Gregory,[46] and Edward Ford,[47] do not consider tectonics as a process. They examine the tectonics of exemplary architectural works by presenting their details as completed objects, rather than as procedures involving various contributors.

The number of craftsmen has decreased considerably in the twenty-first century and most building production no longer involves craftsmen. There are significant differences between the craftsmen of the Arts and Crafts Movement, the Bauhaus Movement, and contemporary contractors, builders, foremen, technicians, and workers. The concept of craftsman can be better understood through Martin Heidegger's book *Being and Time*[48] and Richard Sennett's *The Craftsman*.[49] Contemporary contractors, builders, foremen, technicians, and workers usually have a less specialized skill set, utilize more time-efficient standardized methods,

---

43 Ingberman, *ABC: International Constructivist Architecture*.
44 Bayer, *Bauhaus 1919–1928*.
45 Schwartz, *Introducing Architectural Tectonics*.
46 Gregory, *Comprehensive Tectonics*.
47 Ford, *Five Houses, Ten Details*.
48 Martin Heidegger, *Being and Time*. Trans. *John Macquarrie and Edward Robinson*, 25th Edition. (Blackwell Publishing, 2005[1927]).
49 Richard Sennett, *The Craftsman*. (Yale University Press, 2009).

work with a wider range of materials and systems and employ more advanced technological tools. Their environmental impact also differs. The main distinction between them and earlier craftsmen lies in the structure of the modern economy and business practices, which influence building activities. However, contemporary contractors, builders, foremen, and technicians are still influential in the building practices because they are typically well-versed in the building market, including its materials and systems and are experienced in their application. Together with professionals, they influence both the process of tectonics and its outcome. Consequently, there is insufficient research on the procedural nature of tectonics, as this would require studying the contributions of all actors involved in architecture.

Aim and Claim of this Book

The aim of this book is to provide an in-depth examination of the building process of a house, the changes that occur during this process, their relationship with tectonic affects, and innovative attitudes and the actors of these changes. The house, referred to as the Monarga House in this book, is the author's own house and the author is also the architect of the house. Figure 1.1 displays the east façade of the house, located in the fishermen's village of Boğaztepe (also known as Monarga) in İskele (formerly known as Trikomo), North Cyprus. Chapter 2 of this book gives detailed information about the context of the house.

The term "in-depth examination" includes both the objective and subjective dimensions of the changes that occurred during the building process of the Monarga House. What are these changes during the building process? How and where do they occur? Are they the outcomes of innovative attitudes? Who are the actors responsible for these changes? To what intensity do they occur and are they interconnected? How do these changes impact the construction quality of the building and the tectonic affects caused by the building? Could the image of the building change significantly due to these changes? Are visible tectonic affects considered more important than the general construction quality of the building? Could tectonic affects potentially hide undesirable aspects of the building? Is it possible to map the process of tectonic affects of the Monarga House? Can knowledge about the changes that cause tectonic affects during the construction of a specific building contribute to the theory of tectonics and tectonic affects? Do we underestimate the innovative attitudes that emerge during the construction of simple buildings? Do we underestimate the contributions of contractors, sub-contractors, builders, foremen, technicians, and carpenters who are part of the building team? Could this knowledge benefit the management of architectural, engineering, and construction firms? Is there a need for a procedural theory of tectonics? The RONAC believes that a detailed in-depth examination of this subject can provide a holistic view of tectonics as a process and lead to further inquiries.

The hypothesis of the book is that "tectonics is or can be a process." The book claims that the tectonics of buildings can change considerably during their building processes. However, for these changes to be recognized and made beneficial, a supportive procedural theory of tectonics is needed.

Methodology

The RONAC (Researcher, Owner, Neighbor, Architect, Controller) believes that conducting typical research through questionnaires or interviews with individuals involved in building processes may not lead to an in-depth understanding of this subject. There are several reasons for this. First, it is necessary to capture all changes, tectonic affects, innovative attitudes, and their actors during the building processes, including the application project, tendering process, and construction process of multiple buildings. Some changes may not be noticeable after construction, even when the application project is compared to the completed building. Many changes will be quickly forgotten by all participants. Contractors, sub-contractors, builders, foremen, technicians, and carpenters do not attach significance to many of the changes they make and even important changes may be forgotten immediately. Moreover, many changes may lose their significance over time and may evolve over time as well. Second, tectonic affects are typically subconscious and people's likes and dislikes are usually influenced by their taste and world view.[50]

To provide an in-depth examination of the tectonic process, this research employs a case study methodology[51] through abduction,[52] supported by three in-depth interviews with professionals closely involved in a building's process (OAC, Owner, Architect, Controller; ONACC, Owner, Neighbor, Architect, Controller, Contractor; and OCCC, Owner, Civil Engineer, Controller, Contractor) and four interviews with purposefully selected local professionals, including an experienced contractor, two experienced controllers (one architect and one civil engineer), and the head of the Chamber of Architects in North Cyprus.

---

50 Baker, *From Opinions to Images.*

51 Rolf Johanson, "On Case Study Methodology." In *Methodologies in Housing Research*, Eds: D. U. Vestbro, Y. Hurol and N. Wilkinson. (Urban International Press, 2005): 30–39.

52 Frank Conaty, "Abduction as a Methodological Approach to Case Study Research in Management Accounting – An Illustrative Case" *Accounting, Finance & Governance Review* 27, accessed October 6, 2023, 10.52399/001c.22171; Anna Dubois and Lars-Eric Gadde "Systematic Combining: An Abductive Approach to Case Research" *Journal of Business Research* 55, 2002: 553–560.

No interviews were conducted with foremen, technicians, carpenters, and workers because such interviews need to be conducted in real time and on site to be informative.

Case study methodology is preferred when it is necessary to capture the complexity of a single case. The focus is on understanding that case in all its aspects. This method is widely used in many disciplines, including architecture. The case must be studied in its natural context, often using a variety of methods (qualitative and mixed methods are more common). Incorporating triangulation is also necessary to support the same findings through different approaches. Triangulation can involve using multiple methods, or combining data, theory, and investigations. This research combines a chain of data about changes, their tectonic affects, innovative attitudes, and their relationship to theory to enhance the validity and reliability of the study.[53]

There are five types of case studies. These are Critical, Unique, Typical, Revelatory, and Longitudinal.[54] This research presents a unique case, with its uniqueness arising not from the building itself but from the presence of the RONAC, who can observe every detail of the building process.

Analytical generalizations (as opposed to statistical ones) are possible through case study methodology and there are three primary ways to achieve generalizations as research outcomes.[55] These are deduction, induction, and abduction.[56] Deduction provides analytical generalizations by applying various logical techniques for hypothesis testing, which can lead to theory development. This approach involves comparing the case with existing theories.[57] Induction, on the other hand, is based on field study data and uses that data to develop theory. It begins with theory, moves into empirical data collection, and concludes with theory. Both deduction and induction can be used to develop arguments that result in analytical generalizations.[58] Abduction typically combines elements of both deduction and induction, allowing for hypothesis testing, while also facilitating theory development through inductive data formation.[59] Scientific investigations

---

53 Robert K Yin, *Case Study Research and Applications- Design and Methods.* 6th Edition. (Sage, 2018).

54 Fernandes R. C. Souza Case Studies as Method for Architectural Research, 2015. https://www.researchgate.net/publication/314147521_Case_Studies_as_method_for_architectural_research?channel=doi&linkId=58b708fb92851c471d47a326&showFulltext=true (accessed 16 February 2025).

55 Souza, Case Studies as Method for Architectural Research.

56 Johanson, "On Case Study Methodology."

57 Asvoll Havard "Abduction, Deduction and Induction: Can these Concepts be used for an Understanding of Methodological Processes in Interpretative Case Studies?" *International Journal of Qualitative Studies in Education* 27, no. 3: 289–307.

58 Edward Schippa and John P Nordin, *Keeping Faith with Reason – A Theory of Practical Argumentation.* (Pearson, 2013).

59 Johanson, "On Case Study Methodology."

1 Introduction

through case studies often utilize deduction, induction, and abduction together, though in varying sequences. These different combinations can lead to both theory testing and theory building. The relationship between data and theory, known as syllogism, plays an important role in both testing and building theories through case studies.[60] The seven supportive interviews conducted in this research also contributed to both theory testing and theory development.

Abduction is preferred when there is an unexpected fact about a case.[61] Deep engagement of the researcher with the phenomenon, the location and the related theory is also necessary for utilizing abduction effectively. The incorporation of the RONAC into the process of the Monarga House creates a rare and unique case, as the perspectives of RONAC are all combined in one person. The RONAC played a significant role in the construction of the building alongside another main controller. She was able to collect detailed visual and written data on all aspects of the processes of Monarga House. She used various techniques, including taking observation notes (a diary) and systematically photographing the process. Since the RONAC's academic research area is tectonics, she could delve into the details of the process and develop the necessary objective and subjective interpretations about the quality of the tectonic process of the building. Such detailed data collection would not have been possible through any other methodology. Conducting such detailed research on the tectonic process of a single building has its merits.

Listing some results of this book here helps to clarify the preference for using case study methodology through abduction. During the design and construction process of the small 140 m^2 Monarga House, there were 174 changes, with 15 sets of changes occurring as multiple, interrelated changes triggering one another. Ninety-one of these changes led to tectonic affects, 42 presented opportunities for innovative approaches, and 29 resulted in actual innovative attitudes. The RONAC believes that without a single observer taking multiple perspectives on a building, such detailed records could not have been collected. This research has turned a rare situation into an opportunity.

The research results in proving a hypothesis and constructing theory through the case study. Abduction is particularly useful for theory development because the related theoretical framework, empirical field study, and case analysis develop simultaneously.[62] There are two types of abduction: creative abduction and selective abduction. Creative abduction develops new theories by using

---

60 Havard, "Abduction, Deduction and Induction."
61 Johanson, "On Case Study Methodology."
62 Johanson, "On Case Study Methodology"; Conaty, "Abduction as a Methodological Approach to Case Study Research"; Dubois and Gadde, "Systematic Combining: An Abductive Approach to Case Research."

Methodology | 23

induction to generalize new hypotheses. Knowledge is generated through a process of explaining observed events and demonstrating their relevance,[63] leading to the development of new ideas and concepts. Selective abduction, on the other hand, tests existing theories to identify the most appropriate one for the case. Researchers usually cycle through different techniques multiple times to articulate their theories.[64] This research on the Monarga House also combines theoretical data and inductive data in the form of various merged layers, ultimately proving a hypothesis analytically through induction and contributing to existing theories accordingly.

The research combines an objective approach, based on inductive listing and the subsequent analysis of these lists. A subjective approach complements the objective one, encompassing artistic and aesthetic dimensions and meanings. However, the methodology remains analytical. These approaches are both integral to understanding the tectonic qualities. It is essential to analyze and document changes that are purely technical or technological in nature. Furthermore, it is crucial to describe the impact of contextual factors (including political and economic characteristics) on the entire building process. Additionally, an examination of design decisions and ideas in comparison with the tectonics of the completed building is necessary. The study also requires an exploration of the emotional and affective aspects related to tectonics. Integrating theoretical reflections into the work is vital for the development of alternative tectonics theories through this research. This research can be considered a holistic approach, covering both analytical and technical aspects, as well as artistic and aesthetic considerations. It includes both inductive and deductive techniques within the case study methodology of abduction. Ultimately, this research methodology seeks to bridge the gap between the rational and irrational dimensions of the tectonic process.[65] It proves a hypothesis about the procedural nature of tectonics in architecture. It develops a basis for a procedural theory of tectonics with the help of inductively collected pieces of theoretical reflections within the book.

The additional research, conducted through seven in-depth interviews with specific structured questions asked to professionals with various experiences, helped in proving the hypothesis (three interviews with OAC, ONACC, and OCCC) and developing different perspectives on the changes within building processes (four interviews with experienced professionals). These interviews were

---

63 Helmut Prendinger and Mitsurti Ishizuka "A Creative Abduction Approach to Scientific and Knowledge Discovery" *Knowledge Based Systems* 18, 2005: 321–326.
64 Helmut Prendinger and Mitsurti Ishizuka, "A Creative Abduction Approach."
65 Manuel Delanda, *Intensive Science and Virtual Philosophy.* 9th Edition. (Bloomsbury Academic, 2013[2002]).

1 Introduction

audio recorded and detailed notes were taken during the sessions. The evaluation of these interviews was carried out by logically analyzing the interview notes and listening to audio records whenever needed.

The RONAC initially thought about the idea of crafting this book in the form of a novel, similar to certain qualitative research works.[66] Another alternative research methodology considered for this study was autoethnographic research, which typically results in narrative-style texts.[67] However, these methods were not chosen due to the substantial amount of detail within this research. A good story cannot be told with an overwhelming focus on technical details. The building process underwent numerous (174) changes and some caused tectonic affects and innovative attitudes. They were realized by various actors. While some of these changes were interconnected, most were unrelated. Such content can only be listed and explained inductively. Consequently, presenting this information in the form of storytelling while simultaneously substantiating a hypothesis proved is not feasible.

Description of Data

The RONAC maintained records of all drawings and documents throughout every phase of the project. Over the course of approximately three years of the building process, she also kept a detailed diary (forming observation notes) about the production of the house. This diary has evolved into a text of 149 pages. Furthermore, she extensively documented the process through photography, classifying the photos into 514 categories. Each category contains an average of five photos. There were also seven interviews conducted with professionals to develop arguments between different perspectives.

The RONAC originally intended to collect data for researching the construction of her house, with a primary focus on recording the tectonic qualities of her house. Initially, she did not anticipate changes between her project and the actual construction. However, as changes began to occur, she carefully recorded them. As a result, both the diary and the photos encompassed analytical and technical and tectonic and aesthetic aspects pertaining to the Monarga House's building process.

---

66 Aneta Pavlenko "Autobiographic Narratives as Data in Applied Linguistics" *Applied Linguistics* 28, no. 2, 2007: 163–188.
67 Carolyn Ellis, *The Ethnographic I – A methodological Novel about Autoethnography.* (Rowman & Littlefield, 2003).

Process of Data Analysis

In accordance with the case study protocol,[68] the data analysis process encompassed several steps,[69] which are presented in Chapters 3–7, each dedicated to a different phase of the building process. These analyses form inductive data to prove the hypothesis. The following actions were carried out for this purpose.

- Thousands of categorized photos were analyzed and around 100 photos were selected to represent specific construction phases of the house. These photos were ordered and placed into the corresponding chapters. While working on these photos, changes that the RONAC remembered were recorded and included in the relevant chapters.
- The diary was thoroughly read and edited to identify additional changes in the building process. These changes were also recorded and incorporated into the respective chapters.
- The diary text was coded to complete the changes occurring throughout the building process. This coding was carried out using concepts like *change, idea, problem, contribution,* and *trouble.* These codes were used to locate relevant text segments. Coding facilitated the identification of issues that led to numerous changes and served as a reminder of some forgotten changes. These changes were subsequently added to the appropriate chapters.
- An inductive list of changes for each phase of the building process (2 + 20 phases were presented in Chapters 3–7) was compiled based on the coding results. These changes were then reordered according to the timing of their occurrence to ensure a coherent narrative.[70] The changes that were, and could have been, implemented specifically by builders, foremen, technicians, and carpenters were also identified.
- Changes that impacted (or could have potentially impacted) tectonic affects were singled out and presented in separate tables for each phase of the building process. A change was identified as a tectonic affect if it was visible and had a positive or negative impact on design ideas, according to the RONAC.

---

68 Yin, *Case Study Research and Applications.*
69 A good source for qualitative analysis is Anselm Strauss' book. This book suggests not to carry on the analysis in a mechanical way. Anselm L Strauss, *Qualitative Analysis for Social Scientists.* 14th Edition. (Cambridge University Press, 2003).
70 The term *building process* is used to refer to the entire sequence of activities involved in the design of a building, including the application project, the tendering process, and the construction process. In contrast, the term *construction process* specifically refers to the phase of construction and does not include the design or tendering stages.

1 Introduction

- Changes that might involve innovative attitudes were identified and presented after the presentation of changes and tectonic affects for each phase of the building process. The code "contribution" was used for this purpose, but coding alone was insufficient to capture the innovative attitudes within different phases. This was because nobody was consciously aware that they were being innovative during the building processes and clear codes related to innovation were absent from the diary. Critical reading of the lists of changes and the diary was instrumental in recognizing such modest innovative attitudes. In particular, critical reading of the lists of changes was useful in identifying innovative attitudes that required knowledge and experience within each phase.
- The diary text was coded to ascertain the feelings of the RONAC in relation to various building processes. Codes used for this purpose included *feel, happy, worry, sad, excited, angry, love, shock, dream,* and *graveyard*. These codes were applied to locate pertinent text segments, which were then integrated into the appropriate sections of the book. As affects cause positive or negative feelings, identification of feelings and emotions was necessary. The diary contained numerous statements with the word "worry." The RONAC believes that these statements predominantly represent the perspective of the controller within the RONAC.
- The list of changes at each stage of the building process was analyzed to identify changes which could have been implemented by builders and foremen without consulting professionals. These changes are marked with hashtags (#).

Additional qualitative analysis was conducted to complete the other chapters of the book. Chapter 2, focusing on the context of the building, holds particular significance in in-depth qualitative research, especially regarding the construction process. It serves to support and clarify the background for the content in subsequent chapters. The data analysis for this purpose was carried out as follows:

- Context analysis was performed by qualitatively analyzing the diary text using coding. The codes applied included were *Cyprus, Monarga, heat, neighbor, site, tree, price and cost, dirt, politics, legal, steel,* and *bankrupt*. Extracts from the text corresponding to these codes were gathered and subsequently reordered to form the first layer of Chapter 2.
- Relevant extracts containing the feelings of the RONAC were incorporated into this chapter as a second layer.

A third analysis was undertaken to complete the first part of Chapter 3, which addresses the preliminary design process of the Monarga House. This chapter offered information about the building, explained key design decisions and ideas, discussed its tectonic characteristics and presented changes in the application project phase and the tendering process. In this book, the application project

Process of Data Analysis | 27

phase and tendering process are treated as distinct phases of the building process. The analysis of data to identify design decisions and ideas was conducted as follows:

- The diary text was coded to unearth the specifics of the preliminary design decision and ideas related to the Monarga House. The codes employed in this process were *mother, image, picture, feel, freedom, style, color, four directions, tectonics, domestic,* and *minor architecture.* Some of these design ideas were illustrated using photos of the completed building.
- The relevant extracts containing the feelings of the RONAC about the preliminary design project were integrated into this chapter as a second layer.

Analysis of the changes during the tendering process, their impact on tectonic affects and their innovative character did not necessitate any specialized coding because they were together in a certain place in the diary text. Except the introduction and conclusion of the book, all chapters include related theoretical reflections from various theories, including tectonics, and architectural theories, philosophies (including ethics, aesthetics, and politics), theories on sustainability, construction, structural systems, and anthropology. These theoretical reflections form an additional inductive layer that contributed to theory building at the conclusion of the research.

The inductive information – including changes, tectonic affects, innovative attitudes, the RONAC's feelings and theoretical reflections – provides validity through triangulation by incorporating five parallel layers. These layers, which were collected separately, combine both objective and subjective dimensions. While the objective and subjective aspects complement each other, the RONAC chose not to establish continuous transitions between these texts. Maintaining their separateness within the text facilitates the individual examination of changes, tectonic affects, innovative attitudes, the RONAC's feelings about them, and related theoretical reflections. These components mutually support and encourage the asking of various questions. Chapters 3–7 conclude with maps of the tectonic processes for each phase, tables detailing the types of changes, tectonic affects, innovative attitudes, and their actors, as well as an outline of the theoretical reflections discussed within each chapter. Although Chapters 2–7 do not present generalized results, this type of inductive introduction to the theoretical arguments[71] in the conclusion forms a persuasive character that minimizes the potential for bias in this type of research.

The final maps of tectonic affects and innovative attitudes throughout the processes of the Monarga House is presented, analyzed, and discussed in detail in Chapter 8. This chapter also reveals all the research results, including discussions

---

71 Schippa and Nordin, *Keeping Faith with Reason.*

on the impact of changes on the building's tectonics, the innovative attitudes, the actors responsible for the changes, and an outline of the theoretical reflections from the previous chapters.

The three interviews with the OAK, ONACC, and OCCC, focused on their experiences with the changes that occurred during the process of a specific building, which they closely observed. The interviews explored the tectonic affects caused by these changes, the innovative attitudes they adopted and the key actors involved in these changes. The four interviews with professionals primarily addressed their perspectives on the procedural nature of tectonics whether they believe recognizing the contributions of contractors, builders, foremen, and technicians could be beneficial or not. The results of these interviews are presented in Chapter 8, where the hypothesis is confirmed and in Chapter 9, where the contributions of this book to tectonics theory are discussed.

The objective of the conclusion is twofold: first, to prove and discuss the hypothesis that "tectonics is and can be a process," and second, to develop a correlated tectonics theory with the support of the inductive data and theoretical reflections contained within the book. This theory primarily emphasizes the need to enhance the contributions of contractors, sub-contractors, builders, foremen, and technicians as well as professionals, to the tectonic processes of buildings. The chapter concludes by raising questions about the ideological impact of the increasing materialization of building processes, opening the discussion for potential change.

Drawing Process Maps for Tectonic Affects

There are interrelations between the process of changes, tectonic affects, and innovative attitudes and it is good to view these interrelations within a holistic framework. The best tool to bring these together is graphical and these issues can be conceptualized as two maps. However, these will be process maps rather than representations of physical environments.

Tectonic affects are similar to emotions, and there are emotional maps in academic literature, but these maps do not convey processes. Graphical expressions that illustrate activities in construction processes are available, typically in the form of flow charts that depict different items at different points in time. Some flow charts may even use different rectangular color zones. However, given the need for interrelated zones in the process maps of tectonic affects and innovative attitudes, a more effective approach is to design the map in stages, step by step.

The process maps of tectonic affects and innovative attitudes should have a base which demonstrates the building design and process of the Monarga House with respect to the occurrence of its phases in time. Categories of the building process

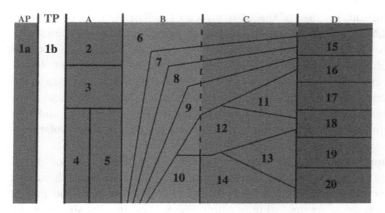

Figure 1.3 The process map without changes. Application project (AP, 1a) and tendering process (TP, 1b), construction of foundations (2), the frame system (3), walls and openings (4), the roof (5), plastering and painting (6), the electrical system (7), the mechanical system (8), ceramics (9), water isolation (10), the windows, shutters, and front door (11), the interior doors (12), the wardrobes and cupboards (13), the fireplace and chimney (14), the garden walls (15), drainage (16), the garage and concrete work in the garden (17), metalwork (18), the pathways (19), and timber work in the garden (20).

in this book have five basic phases which are presented in Figure 1.3 with different letters and different tones of gray.

These five phases are:

- The application project phase (AP) is shown as 1a. The other phases cannot start before the application project is accepted and approved. That is the reason behind showing it as a totally separate vertical column. Similarly, the tendering process (TP) which also takes place before construction, is shown with 1b and as a vertical white column.
- The early steps of construction (Phase A) including the construction of the foundations, structural system, infill walls, heat isolation, openings, roof, and pergolas which are studied in Chapter 4, are presented with numbers from 2 to 5. First, the foundations were built. Once they were complete, the frame system was constructed. After the frame system, the construction of the walls, openings, insulation, and roof system followed. Because of this, the last two items within this phase (4 and 5) are shown as simultaneous (as parallel vertical columns).
- The phase containing plastering and painting, electrical systems, mechanical systems, ceramics, and water isolation (Phase B) which were presented in Chapter 5, is shown with numbers between 6 and 10. The construction process of most of these items started after the early steps of construction and extended towards the later phases. The plastering and painting item extended to the end

of construction. Because of this, it is presented differently than other items in this phase. The application of electrical systems, mechanical systems, and ceramics ended before construction of items in the garden started. These extensions of phases are represented by diagonal lines which reach the later phases. Water isolation is an essential part of many sections of the building such as the foundations and roof. This book classifies the water isolation in the foundations and roof as essential parts of those items. Only the remaining water isolation work is studied separately within this phase. Because of this, water isolation is shown as taking place only in Phase B.

- The phase containing the windows, shutters, front door, interior doors, cupboards and wardrobes, fireplace, and chimney (Phase C), which is studied in Chapter 6, is shown with numbers between 11 and 14. Except for the interior doors, these items are applied only within Phase C as they are shown in the Figure 8.1. Since the door frames and leaves were produced and placed within Phases B and C in the Monarga House, they are shown like that.
- The last section in this figure presents the construction process of the items in the garden (Phase D). These include the garden walls, drainage, concrete work in the garden, garage, fences, dog kennels, garden gates, pathways, roads, and timberwork in the garden. Construction of these elements also follow on from each other. Except for the extension of plastering and painting into this phase, the elements within this phase also had a linear downward process direction.

The foundational process map for tectonic affects may vary from building to building. Therefore, it is essential to prepare a separate process base for different buildings that use varying materials and systems.

The changes that cause tectonic affects and innovative attitudes are marked on two separate base maps accordingly. If a change occurs in Phase B, stage 7, it must be marked at that specific point on the map. Different types of changes with tectonic affects are symbolized differently: eliminated changes (unwanted changes that were stopped are shown as circles in Table 1.1), singular changes (which appear only once and are not related to other changes are shown as hexagons) and multiple changes (which trigger or are triggered by other changes are shown as triangles). Each symbol can be white, gray, or black depending on the type of tectonic affect it causes. White represents positive affects, black represents negative affects and gray represents neutral affects. Table 1.1 presents the symbols used for various changes that cause different tectonic affects.

Eliminated changes and singular changes stand alone on the map. However, multiple changes should be connected to one another to visualize the set of changes together. For this purpose, the related multiple changes are linked with lines.

Table 1.1 Symbols for the types of changes causing tectonic affects.

The symbol	Definition
	Eliminated changes
	Singular changes
	Multiple changes

Lighter colors show the positive character of the change.

The Ways of Avoiding Bias

To ensure the integrity of this research, which is mainly focused on a single building and a single person, it is imperative to clarify and substantiate the means used to avoid bias. Three primary methods for avoiding bias in this abductive case study research are as follows:

1) *Transparency of Researcher's (the RONAC's) identity and relationship to the house:* It is crucial to classify the identity and relationship of the RONAC to this study. The RONAC is an academic architect who specializes in teaching tectonics and conducting research in architectural tectonics. The Monarga House, the subject of this research, is her own design as both the architect and the owner. This dual role might potentially introduce bias as it could lead to unacceptable praise for the Monarga House. Additionally, the RONAC, as the architect and owner, may naturally possess a strong affinity for the architecture of the house she designed. Furthermore, she may be inclined to commend her contributions to the tectonic process of the house.

   A potential source of bias may also stem from the issues with the initial contractor of the house. These challenges could lead to a subjective perspective regarding changes that happened before and after parting with this contractor. To mitigate these biases, rigorous review and reevaluation of the research results and their theoretical contributions against these potential sources of bias, as well as subsequent revisions to problematic text sections, were undertaken. Detailed description of the context of the Monarga House was also provided for this purpose.

2) *Inductive character of intermediate chapters:* The second and most significant method for bias avoidance in this research is the inductive nature of the intermediate chapters of the book. The information presented in these chapters serves to establish and prove the research hypothesis, deriving results methodically in Chapter 8. Changes implemented during each phase of the building process, their tectonic affects, and the associated innovative attitudes are systematically and inductively listed in Chapters 3–7. Information provided about the changes and innovative attitudes is objective. While the information regarding tectonic affects of changes involves a level of subjectivity, this bias is mitigated by the RONAC's combined roles as both the architect and owner of the Monarga House. As the architect and owner of the house, her preferences can be determining. This harmony between roles is conductive to an accurate assessment of the tectonic character, as architects and owners are naturally suited for this evaluation. Chapter 8 compiles and presents the results from Chapters 3–7 as an inductive process map, facilitating the verification of the hypothesis and derivation of results in a systematic manner.

It is important to note that the RONAC's initial stance was against changes during construction, believing that a contract with the contractor and the existence of a list of selected items, materials, and systems in the agreement should require strict adherence to the application project. She was open to changes only if all parties involved, including the contractor, the owner, and the main controller (an experienced civil engineer), unanimously accepted such changes. However, the realities of the building process led to a change in this perspective. The RONAC subsequently discussed the challenges associated with changes in construction with her academic colleagues. This prompted her to carefully document the changes in her diary, encompassing all tectonic aspects. She also decided to photograph each phase, with particular emphasis on capturing changes, both before and after they occurred. This research consequently shifted its focus to the changes that occurred in the Monarga House.

3) *Inclusion of the RONAC's feelings:* The presentation of the RONAC's feelings within the text serves as a mechanism to highlight how bias may exist. This openness contributes to maintaining transparency and integrity in the research.

4) *Conducting seven interviews:* These interviews with three professionals who followed the building closely and four other interviews with experienced

professionals were carried out to verify the research findings and to develop diverse perspectives that fostered discussion.

Contributions of this Book

The contributions of this book to tectonics theory are based on the successful demonstration of the hypothesis that "tectonics is or can be a process." This achievement facilitates the development of the core principles of a tectonics theory that advocates more comprehensive and collaborative building processes compared to existing collaboration models within the building processes. This approach to tectonics can potentially reduce the procedural instrumentalization of buildings and optimize opportunities for enhancing tectonic qualities through the active inclusion of contractors and sub-contractors, builders, and foremen in the process of continuous design evolution during all building processes. The theory recognizes the contributions of these stakeholders to current building processes and advocates for an inclusive building process to leverage their expertise and experience. In doing so, this theory contributes to existing tectonics theories, clarifies the impact of concealed building components and tectonic qualities and highlights the influence of errors, details, and camouflage on the tectonics of buildings. The insights provided by this research serve to improve existing theories about details related to errors, camouflage, and concealed building components through an examination of the tectonic process of the Monarga House, unveiling various degrees and types of these concepts within the context of a single building.

The outcomes of this research are expected to stimulate further studies and innovations in the management of building processes, including both office and construction management. This could manifest in the form of an increased integration of digital technologies such as BIM. The theoretical contributions of the book may also initiate changes in building processes led by architectural, engineering, and construction firms by advocating for greater participation and contributions from builders, foremen, technicians, and carpenters in shaping the tectonics of buildings.

While the knowledge presented in this book primarily relates to the process of tectonic affects within a specific building, demonstrating that a procedural approach is possible, creating two process maps (one for the changes and tectonic affects and one for innovative attitudes) can be adapted to the specific needs of construction firms and education in schools of architecture. Such maps can be employed during the processes of individual buildings for various purposes, including presenting the entire process comprehensively and enhancing communication among all stakeholders.

Readership

This book can have a diverse array of potential readers across multiple disciplines and knowledge levels. Here is an overview of its target audience:

1) *Postgraduate students and academics in tectonics, theory of architecture, and construction:* This book's dataset, research findings, and contributions to tectonics theory can serve as a valuable resource for postgraduate students and academics engaged in research in fields related to tectonics, architectural theory, and construction.

2) *Undergraduate students of architecture and civil engineering:* It is possible to read this book selectively by focusing on certain explicitly differentiated layers. These layers include the technical and analytical layer (which covers details and issues related to construction), the tectonic affects layer, the innovative attitudes layer, the feelings layer (representing the various roles of the RONAC) and the theoretical reflections layer, which incorporates various schools of thought and disciplines based on their relevance. The process map drawings might also be of interest to architecture and civil engineering students. The book offers insights into the construction process through photographs, related concepts, and explanations. It provides students with a realistic look at the complexities of building processes. It also presents a holistic perspective of the architectural and engineering professions, encompassing the roles of an architect as a designer, controller, and even a contractor. Students interested in becoming controllers or contractors in the future may find the actual conditions presented in this book particularly relevant. The book bridges the technical, technological, and analytical dimension with the artistic, aesthetic, and meaningful dimension and imparts an understanding of the values architects and civil engineers uphold and the challenges they face. Scholars, especially those seeking to reform the way construction courses are taught in architecture schools, may find this book useful as a reference to explore alternative pedagogical approaches.

3) *Architects:* Practicing architects can find valuable insights in the book, potentially offering them new perspectives on building processes and the tectonic dimension of their work. *Engineering, construction firms, controllers, and architecture*: For professionals in these domains, this book can be particularly relevant if they are interested in building processes and developing communication techniques between owners, controllers, and contractors,

sub-contractors, builders, foremen, technicians, and carpenters. The insights offered may assist in the development of novel methods for effectively managing change in their activities.

The multifaceted nature of this book allows it to resonate with broad readership and addresses the intersection of architectural, engineering, construction, and tectonic principles.

Part I

Context of the Building, Its Project, and Tectonic Characteristics

Part I contains two chapters: Chapters 2 and 3. Chapter 2 is about the context of the Monarga House and Chapter 3 is about the preliminary design project, application project, and tendering phase of the building process. Chapter 2 informs readers of the political and economic dimensions of the context and provides an in-depth understanding of the site. Chapter 3 presents photos of the building and drawings from the preliminary design project first, then explains the main design ideas, and analyzes the building's tectonic characteristics. This chapter then presents some drawings of the application project, the changes that occurred during the application project, the changes which caused tectonic affects, and the changes which initiated innovative attitudes. The changes that occurred during the tendering process, their tectonic affects, and their relationship with innovative attitudes are also described in this chapter. This part of the book covers the preparatory phases of the building process. Part II will move on to the physical construction of the building and the changes that happened during the construction process.

Tectonics as a Process in Architecture, First Edition. Yonca Hurol.
© 2025 John Wiley & Sons, Inc. Published 2025 by John Wiley & Sons, Inc.

2

Context of the Project and Building and the Early Story of the House

There are detailed and various types of explanations about the context of the Monarga House in the diary (observation notes) of the RONAC (Researcher, Owner, Neighbor, Architect, Controller). The diary was coded using the concepts of *Cyprus, Monarga, heat, neighbor, site, tree, price and cost, dirt, politics, legal, steel* and *bankrupt* in order to catch the attributes related to the context of the house. The matching statements were extracted, reconsidered, and grouped before being used in the text. Later, the diary was searched for feelings about the whole process by using the codes: *feel, happy, worry, sad, excited, angry, love, shock, dream, graveyard, stress,* and *shout*. These extracts about feelings are present in all chapters of the book including Chapter 2. The chapter also contains some theoretical reflections wherever appropriate. Therefore, Chapter 2 consists of three integrated layers: the layer of context, the layer of feelings, and the layer of theoretical reflections.

One of the dominant contextual issues about the Monarga House is the political economic context. The political economy might not be a particularly important issue for many architectural contexts, if the context is within a recognized, democratic, and modern country. If that is the case, then the political context has been determined by social agreements and laws. However, within a contested zone, particular features of the political economy might deeply influence the characteristics of the context.

About the Political Economy of Property Ownership in a Conflict Zone

The Monarga House is in North Cyprus which is an unrecognized country. The north side of the island of Cyprus, located in the Mediterranean Sea, became isolated from much of the world after 1974 due to political turmoil, which led to

Tectonics as a Process in Architecture, First Edition. Yonca Hurol.
© 2025 John Wiley & Sons, Inc. Published 2025 by John Wiley & Sons, Inc.

wars and massacres of Greek and Turkish Cypriots. This isolation made North Cyprus economically dependent on Türkiye.

> *Reflection(s) from Theory*
>
> **A conflict zone** is a place where there is fighting between people. However, the problem might not be resolved even after the fighting ends. According to Robert Bevan, buildings which represent the identity of rival groups might be deliberately damaged during and after wars and conflicts. Since historical buildings form evidence for the historical presence of certain groups of people, they might attract more hostility.[1] The condition of other buildings which do not directly represent their owners' identity is another issue. According to Ceren Boğaç, Turkish Cypriot people who resettled in Greek Cypriot houses have faced some **problems of place attachment.**[2] They have not maintained these buildings for many years, and they have protected the furniture and belongings of the previous owners. Since Cyprus is the longest unresolved conflict zone, it is also perceived as having an **uncanny geography** which can even act as an emotional nexus between people and this strange environment.[3]
>
> If a conflict is not resolved, its environment becomes a **contested environment**. Socrates Stratis wrote about the contested environments in Cyprus and discussed many projects to move towards building a decisive critical resistance against the dominant nationalist trends with the help of projects like Hands on Famagusta and Contested Fronts.[4] These projects can also be accepted as **transitional justice activities** which reveals the truths of the past.[5] These projects also cover **research on the roots of multiculturality** and the **production of alternative cultural heritage** on the island.[6] There have also been **transgressive design activities**[7] to reflect the dreams of

---

1 Robert Bevan, *The Destruction of Memory – Architecture at War*. 2nd Edition. (University of Chicago Press, 2016).

2 Ceren Boğaç "Place Attachment in a Foreign Settlement" *Journal of Environmental Psychology* 29, no. 2, 2009: 267–278.

3 Ceren Boğaç "The Process of Developing an Emotional Nexus between the Self and an Uncanny Geography: An Autoenthnography" *Emotion, Space and Society* 36, 2020: 100688.

4 Socrates Striatis, "Contested Fronts: Commoning Practices for Conflict Transformation", 2016, accessed February 17, 2025. https://www.cy-arch.com/contested-fronts-commoning-practices-for-conflict-transformation/.

5 Emily Farell and Kathy Seipp, *The Road to Peace – A Teaching Guide on Local and Global Transitional Justice*. (The Advocates of Human Rights, 2008).

6 P. Nora and L. D. Kritzman (eds.), *Realms of Memory – The Construction of the French Past, Volume 1 – Conflicts and Divisions*. (Columbia University Press, 1996).

7 Bertuğ Özarısoy and Haşim Altan, *Transgressive Design Strategies for Utopian Cities – Theories, Methodologies and Cases in Architecture and Urbanism*. (Routledge, 2023).

> Cypriots in impossible architectural projects as one of the various **symbolic compensation platforms**.[8] Transgressive design enables a continuous shift in the point of view which results in greater freedom during design and related research.[9]
>
> **Nationalism** is a rival of beliefs and thoughts which seek peace. According to Theodor W. Adorno and Max Horkheimer, contemporary nationalism, which is based on the definition of the other(s), originates from enlightenment and secularity. **Identity thinking** has epistemological roots and relates to the object-subject-concept relationship. Enlightenment looks towards freedom; however, because of its epistemological roots, it is contrary to any undefinable issues. This results in the application of power to define all the undefinable ideologically. The subject (self) can be built in this way. The concept of **other** is immanent to the concept of self and it is subjected to legalized power. Therefore, different definitions of self (nationalisms) cause conflicts.[10] Nationalism in North Cyprus also goes hand in hand with protecting seized property.

The inflation and foreign currency crisis in Türkiye relates directly to North Cyprus and caused a continuous increase in land and house prices. Since money is not very reliable, people prefer to invest in land and buildings, and this further increases the economic value of all types of property. The RONAC expresses such issues in her diary as follows:

> I was worried about the political situation in Cyprus. Whether North Cyprus agrees with Greeks or Turks, I thought that the land and house prices will continue to increase and soon this will make it impossible for me to own a reasonable house or flat.
>
> ... The changes in Sterling were also funny. At the beginning it went up then slightly and then dropped because of Brexit. Now it is flying. This works for me. When I turned my retirement savings into Sterling it was 3.7 TL

---

8 Edward W Soja, *Seeking Spatial Justice*. (University of Minnesota Press, 2010). Henri Lefevbre, *The Right to the City (Le droit a la Ville)*. (Verso, 2017[1968]). David Harvey, *Social Justice and the City*. (University of Georgia Press, 2009[1973]). Susan S Fainstein, *The Just City*. (Cornell University Press, 2010).

9 The concept of transgression was initiated by Michel Foucault and reflected the field of architecture by Bernard Tschumi, Leonard Lawlor and John Nale, "Transgression." In *The Cambridge Foucault Lexicon*, Eds: L. Lawlor and J. Nale. (Cambridge University Press, 2015): 509–516. Bernard Tschumi, "Architecture and Transgression." In *Architecture and Disjunction*. (MIT Press, 1996[1976])): 64–78.

10 Theodor W Adorno and Max Horkheimer, *Dialectics of Enlightenment*. Trans. J. Cumming. 4th Impression. (Verso, 1995[1947]).

2 Context of the Project and Building and the Early Story of the House

> (Turkish Lira), and now it is 5.25 TL (in 2018 ... in 2022 it was 23 TL, and in 2024 it is almost 45 TL). I feel safe with Sterling. I hope I will be able to complete my house without any money problems. I hear from many of my friends that their houses' construction took 4 or 5 years. Some of them have not completed yet.

Since the country is isolated, there are few sectors in Cyprus which actively contribute to the economy of the country. These are mainly higher education, tourism, and construction. The majority of construction workers are migrants from Afghanistan, Syria, and African countries. As for the news about human trafficking for labor in the North Cyprus 2024 newspapers, it is indeed a worrying issue.[11] Similarly, there are many newspaper articles about accidents on construction sites due to a lack of safety precautions, inexperienced workers, and insufficient supervision.[12] Foreign students in universities (even master's and PhD students also work in construction). Turkish Cypriot construction workers prefer to work in the Republic of Cyprus to earn more money. There are also many casinos and banks in North Cyprus, but it is difficult to categorize these as productive sectors.

The unrecognized situation of the country has a great effect on the status of land and buildings. Not all land and buildings have the same status, and this originates from the war and resettlement of the first owners of many of these properties on the south side of the island (the Republic of Cyprus, a recognized country). The RONAC explains the status of North Cyprus property in her diary as follows:

> One of my colleagues in our school warned me about the types of title deeds in North Cyprus. There are four main types. The first is the Turkish title deed. The second corresponds to property in the Republic of Cyprus ("eşdeğer" in Turkish). The third refers to points awarded to citizens by the North Cyprus government in relation to the property they left behind. The fourth type of property is originally Greek property, but it is allocated to its new users by the North Cyprus government ("tahsis" in Turkish). My colleague told me to buy either the first or second type and never the other two; especially the last one. I found out that this last type did not even belong to the people who kept the deeds of these houses. As I do not have other resources in life, I decided to use my money in a safe and ethical way. I started to look for a property of the appropriate dimensions and an acceptable title deed.

---

11 Pinar Barut, "Sosyal Medyada Köle Pazarlığı" *Özgür Gazete*, September 4, 2024.

12 Anon, "Alınan Tedbirler Yetersiz (Precautions are Not Sufficient)" *Yenidüzen Gazetesi*, October 20, 2024; Anon, "Nedenler Net; Tedbirlerin Yetersizliği, Çalışanların Bilgi Yetersizliği ve Eksik Denetim (Reasons are Clear: Precautions are not Sufficient, Workers are Inexperienced, and There is Insufficient Oversight)" *Özgür Gazete* October 20, 2024.

Houses with a Turkish title deed belonged to Turkish Cypriot people before the 1974 war. Houses which correspond to property in the Republic of Cyprus, were exchanged after the war. A Turkish Cypriot who had left a house in the new Republic of Cyprus made an official application to exchange his/her house with another house in North Cyprus. The house in North Cyprus belonged to a Greek Cypriot before the war. The North Cyprus government also defined some points relating to the property of Turkish Cypriots who are now in the Republic of Cyprus. Some houses in North Cyprus were exchanged using these points and allocated to Turkish Cypriots. The title deeds of these houses form the third type. The fourth type of title deed is given to the owners without any correspondence. These houses belong to Greek Cypriots, and they are allocated to Turkish Cypriots or Turkish people from Türkiye. The Republic of Cyprus also has some problems with Turkish Cypriot property on their land. However, their situation is different and does not apply to the RONAC's house.

Reflection(s) from Theory

Christopher Morton wrote that **not fighting for the lost home** causes serious harm to people's capacity to fight.[13] The RONAC believes that experiencing this situation is one of the reasons for the never-ending struggle of Greek Cypriots for their lost homes. In October 2020, when Varosha (Maraş in Turkish) was accessible to the public for the first time in 46 years, some Greek Cypriots visited and stood crying in front of their old houses. Many Turkish Cypriots also lost their homes during the 1974 war. Many of these people were resettled in Greek Cypriot houses in North Cyprus. However, it is generally known that this **resettlement** in North Cyprus was not fairly carried out.

The RONAC has lived as a tenant in Monarga, a fishermen's village for 15 years. She conducted research about her previous house and published it.[14] Her research was about the ethics of owning and using such houses which were originally owned by Greek Cypriots or Maronites. Because of this research she learned about the history of Monarga. Greek and Turkish Cypriots lived side by side in the village before the events of 1963 when British colonial rule ended. The natural fortifications of Monarga were used by British army snipers before 1963. Turkish Cypriot properties (usually farms) were demolished and Monarga became Greek

---

13 Christopher Morton "Remembering the House – Memory and Materiality in North Botswana" *Journal of Material Culture* 12, no. 2, 2007: 157–179.
14 Yonca Hurol and Guita Farivarsadri, "Reading Trails and Inscriptions around an Old Bus-house in Monarga, North Cyprus, *Building Walls and Making Borders: Social Imaginaries and the Challenge of Alterity.*" In , Eds: M. Stephenson and L. Z. Aldershot. (Ashgate Publishing, 2013): 155–176.

2 *Context of the Project and Building and the Early Story of the House*

Cypriot in 1963. Then the Greek Cypriots lost their village properties due to the 1974 war. Some properties in the village were given to Turkish Cypriots who were originally from Monarga. However, the remainder were given to other Turkish Cypriots or Turkish people. In 2004 when the borders were opened in parallel with developments related to the Annan Plan which was about the unification of the two sides in Cyprus, many Cypriots on both sides visited their old houses. Many Greek Cypriots visited Monarga and they still do.

Reflection(s) from Theory

Yonca Hurol and Guita Farivarsadri wrote about a Maronite family who left their house in Monarga due to the war in 1974.[15]

> The Maronite family left their house on 14 August to visit one of their relatives in Nicosia. They decided to take their fourteen-year-old daughter, and their seventeen- and eighteen-year-old sons with them at the last moment. However, they forgot their radio, which they had left on the kitchen windowsill. Having a radio was really important for hearing the news, and they thought it was going to be stolen. They were not expecting to encounter any other particular trouble. However, all the roads towards the east of the island closed that afternoon, and all their neighbors abandoned their houses. The Maronite family, therefore, did not see their house again till the borders were opened in 2004. The ground floor of their house was left half finished, while the other half of their house had hardly been started ... After they left, a high-ranking member of staff from the Turkish army was allocated their house, until it was subsequently allocated to a Turkish Cypriot family, who were originally from Monarga.

This extract shows that Turkish and Greek Cypriots left their houses **without knowing they would ever be able to return**. They hadn't made any decision to leave their homes, so they had not made the necessary preparations.

In 2004, many Greek Cypriots found that their houses were almost untouched except for deterioration due to time. The feelings caused by these visits can be better understood through Marcel Proust's novel *In Search of Lost Time*. According to the last volume of this novel, if a person has the same haptic experience later that s/he had during his/her childhood, s/he experiences a flashback of memories. These flashbacks can make people

---

15 Hurol and Farivarsadri. "Reading Trails and Inscriptions."

> aware of the lost potentials of their childhood.[16] Coming back to a childhood home which was left traumatically, finding that home untouched and **experiencing the same haptic affects (especially the errors)** after 40 years provoke flashbacks of many memories and a deep recollection of the pre-trauma childhood. The RONAC heard from her Scottish neighbor that when the Maronite family visited their house for the first time in 2004, they cried in front of the metal shed at the back of the house because they remembered how their deceased uncle had helped them during its construction.

Monarga is divided into four districts. The lower part of the village nearest the sea contains many popular restaurants, some businesses activities, a harbor, and a village square. In the hills of Monarga there are two separate parts looking towards the sea from the top. One of these is inhabited by more affluent people and the other is mostly inhabited by the original Monarga people. There is another part of the village behind that is usually inhabited by Turkish migrants. The RONAC lived in a Maronite's house, in the part inhabited by the original Monarga people. She loved the village and tried to take care of it all the time. This house in front of Monarga's natural fortifications, faces the sea. Then, she bought a plot of land which can be seen from the windows of the Maronite's house, to build the Monarga House. Being a neighbor helped her a lot during the construction work because she heard and saw everything happening at the construction site from her house.

> *Reflection(s) from Theory*
>
> Hurol and Farivarsadri[17] wrote that "The **traces of war** on these buildings can still be seen. The bullet damage to the flat-roofed extensions, and to the brick covered parts of the façades is still visible. A propellor from a crashed Turkish airplane is also kept in the garden." The old lady who was the new inhabitant of the house, also kept her **wartime camping equipment** (cooking and washing equipment etc.) in a dwelling made from an old bus. "Whilst explaining, in her novel, the 1974 re-settlements of the evacuated Greek houses, Nese Yaşın[18] mentions the striking differences between these luxurious Greek houses and their impoverished new owners. In those days the houses still represented their first owners." Greek Cypriots were the affluent group of people on the island before 1974.

---

16 Marcel Proust, *Search of Lost Time,* 7 volumes. (Everyman's Library, 2001[1913–1927]). The titles of the seven volumes are *Swann's Way, In the Shadow of Young Girls in Flower, The Guermantes Way, Sodom and Gomorrah, The Prisoner, The Fugitive,* and *Time Regained.*
17 Hurol and Farivarsadri, "Reading Trails and Inscriptions."
18 Neşe Yaşın, *Üzgün Kızların Gizli Tarihi* (Secret History of the Sad Girls). (İletişim, 2002).

2 Context of the Project and Building and the Early Story of the House

It was obvious that most of the new residents of these buildings did not take care of them because they thought they might have to leave them in the future depending on the political situation in Cyprus. These people usually bought cars instead of investing in land or property.

Reflection(s) from Theory

In parallel with the peace building and transitional justice activities on the island, there was also a peak in construction activities after 2004 because it was suggested in the Annan Plan that if the new users had taken up more than 60% of building activity to change the building/land, they should be accepted as the owners of these properties.[19] Many people built on land which was originally Greek Cypriot and many people made radical changes to their properties.

Most of the original Greek Cypriot houses were examples of modernist architecture. In Deleuze and Guattari's terms, they are examples of **deterritorialization** and it is not possible to see any cultural expressions in them.[20] There were also some **territorialized** spaces mainly around public buildings such as churches, government buildings, schools, theaters, halls, and so on. Greek Cypriots could not find time to construct buildings based on postmodern architecture on the north side of the island. Postmodern architecture is better expressed as **reterritorialization** in Deleuze and Guitar's terms. The majority of new buildings of Turkish Cypriots on Greek Cypriot lands and the redesigned and rebuilt Greek Cypriot buildings can be accepted as examples of reterritorialization which are attempts to create meaning.

The RONAC is a Turkish Cypriot who grew up in Türkiye. She wanted a house in North Cyprus because she became attached to Cyprus during her childhood, through her mother's family and school stories, and a rich collection of photographs kept in two suitcases. These photos were taken by her mother from 1942 onwards when her father bought her a camera.

---

19 International Peace Research Institute, "The Annan Plan for Cyprus – A Citizens Guide," 2003, accessed February 17, 2025, file:///C:/Users/DC/Downloads/ENG%20The%20Annan%20Plan%20for%20Cyprus%20A%20Citizen%E2%80%99s%20Guide.pdf (accessed 17 February 2025).
20 Gilles Deleuze and Felix Guattari, *A Thousand Plateaus, Capitalism and Schizophrenia*. Trans. Brian Massumi. 5th Edition. (Continuum, 2004[1980]).

> **Reflection(s) from Theory**
>
> Cyprus was a multicultural country during the first half of the twentieth century. The RONAC's mother was educated in a Catholic boarding school while her father was educated in a British boarding school. They both had Greek, British, French, and Armenian friends. They were living together and sharing their cultures as equals. However, Cyprus was divided in the twentieth century and **multiculturality** exists in another way. Greek Cypriots and Turkish Cypriots live on separate lands. There is multiculturality in both countries because of migration due to wars, the presence of international universities, and tourism. Both countries are governing nations and they both prefer to write the history of the island as they wish. Yara Saifi compared the two national museums on both sides of Cyprus (the Lefkosa Museum of National Liberation in Lefkosa/Nicosia and the National Liberation Museum in Nicosia/Lefkosa – the divided city) in her PhD thesis about the architectural context of conflicted spaces.[21] Both museums represent their nationalistic approaches to history. As Pierre Nora describes "what we are now in the habit of calling 'memory' is in reality the history of those who have been forgotten by history, those who have been excluded from **official history** because they live on the margins of society."[22]

The houses in Monarga have various title deeds. Some of them are the second type and others are the third or fourth type. The RONAC has not seen any houses with a Turkish title deed in Monarga.

The status of land in North Cyprus also changes according to some regulations. Some land can be used according to "20 + 15" and some can be used according to "Fasıl 96." Fasıl 96 allows apartment blocks on the land, while "20 + 15" allows only small buildings. Fasıl 96 sites are larger, and it is possible to use a larger portion of the site. Although the RONAC thought this would be an opportunity to demolish the house and build an apartment, she decided against having apartments built around her house in the future. She decided not to go in this direction when buying her land.

The RONAC was also warned about some other property conditions. As Cyprus has a very rich history, there are some historical sites which could yield some valuable archaeological remains that would eventually stop construction. People believe that initial surveys relating to the historical value of the site are not

---

21 Yara Saifi, On Political Conflict and Architecture – Evaluation of the Architectural Context of Jerusalem's Conflict. (PhD Thesis, Eastern Mediterranean University, 2012).
22 Nora and Kritzman (Eds.), *Realms of Memory*.

2 Context of the Project and Building and the Early Story of the House

sufficiently reliable. The RONAC was also warned about land which is below pavement level; this increases the cost of foundations due to drainage problems. The RONAC thinks that these issues discourage people from buying property or building.

There are also some tales about the cost of a house. One relates to the cost of the house per square meter. In 2016, the RONAC was told that a house like hers would cost £500 sterling per square meter. However, this was not true for several reasons. The soil type and roof type, the use of a special type of plaster, the presence of pergolas, shutters, and a fireplace, and a preference for certain materials and systems, and so on, increase the cost per square meter. However, if such features were removed, the house would lose a lot of its character.

The RONAC had cost calculations done by a contractor as well as by the architectural firm which dealt with the application project. The results were higher than expected. So, she started to think about how the cost could be decreased without harming the project. Reasons for the high costs were the rocky soil (which is safer during earthquakes), the pitched roof (preferred for comfort in hot weather), the covered terraces, the special plaster (which is twice the price of normal plaster because of the need for special workmanship and hand painting), the shutters (needed for security), the use of heat isolated mullions and double glazing for openings (to retain heat), the wardrobes, cupboards, and a closed garage which increased the area of the house by more than $20 \, m^2$.

One construction firm suggested that the RONAC could lower the total cost by not having a controller. This was a strange idea and the architect of the application project was also against this.

Reflection(s) from Theory

According to Ian McKinnon, signs of **corruption in the construction sector** include a "refusal to deal with anyone other than the main point of contact." Many people in the building sector do not want to discuss their roles with professionals. Indications of corruption in the construction sector are listed by Ian McKinnon as:

- Unnecessary, inappropriate, and poor-quality goods or services[23]
- Suspicious invoices
- Reduced commitment to quality, ethics, and compliance

---

23 Ian McKinnon, Corruption in Construction: How to Tackle this Industry-wide Problem, 2020, accessed February 18, 2025. https://www.chas.co.uk/blog/tackling-corruption-in-construction/.

> - Biased procurement procedures
> - **The supplier refuses to deal with anyone other than their main point of contact**
> - Individuals are reluctant to take annual leave or offer insight into their role
> - A sudden increase in work opportunities
> - A history of corruption
> - Absence of written agreements.
>
> According to Albert Chan and Emmanuel Owusu, forms of corruption in the construction industry are: bribery, fraud (falsification), collusion, embezzlement, nepotism, extortion, conflict of interest, big rigging, kick-backs, professional negligence, front shell companies, favoritism, cronyism, dishonesty, facilitation payments, price fixing, guanxi, patronage, client abuse, clientelism, ghosting, influence peddling, money laundering, lobbying, intimidation and threats, coercion, cartels, blackmail, solicitation and deception. The most common forms of corruption are bribery, fraud (falsification), and collusion.[24]
>
> According to the Global Economic Crime Survey, the most serious economic crimes are committed in the engineering and construction industry (70%). Fraud risk assessment is suggested to avoid such economic crimes.[25] The sixteenth of the 17 Sustainable Development Goals (SDGs) for architecture within the United Nations Guidebook also suggests peace, justice, and strong institutions which can **avoid all types of corruption within the construction industry**.[26]

Similar to Türkiye, some North Cyprus contractors are too involved in both construction and politics. Because of this, they usually cannot plan their time and resources. They obtain jobs which they cannot complete on time. They also use their customers' money for various political purposes which lead to bankruptcy. If the contractors are too busy, they expect the owner and controller to tell the workers what to do. This requires an awareness of the situation and great care to eliminate problems.

---

24 Albert P.C. Chan et al., "Corruption Forms in the Construction Industry: Literature Review" *Journal of Construction and Engineering Management* 143, no. 8, 2017: 04017057.

25 PwC's 2014 Global Economic Crime Survey, Fighting Corruption and Bribery in the Construction Industry, accessed December 18, 2022. https://www.pwc.com/gx/en/economic-crime-survey/assets/economic-crime-survey-2014-construction.pdf.

26 United Nations, An Architecture Guide to the UN 17 Sustainable Development Goals, 2018, accessed February 16, 2025. https://www.uia-architectes.org/wp-content/uploads/2022/03/sdg_commission_un17_guidebook.pdf.

Context of Building Design and Construction

With a limited budget, the RONAC had to search though all possibilities of buying a house. First, she looked for a ready-made house because she thought it was risky to have a house built.

> My first intention was to buy a house with a garden where I could continue to take care of my pets. I thought it was going to be too stressful to design a house and have it built. I also heard that quality decreases in those cases. I also heard quite a lot of negative stories. Although my closest friend had designed and built his own house quite recently, I decided not to do the same thing.

However, she refused to consider any of the ready-made houses, mainly because of title deed problems, their site dimensions, their relationship to other buildings and roads, their badly designed staircases, and the fact that they were too large and prestigious. The RONAC almost bought a large and expensive house with two huge symbolic columns at its entrance, particularly because its staircase was easier to climb, but her best friend said, "Yonca this is not you!" and she changed her mind.

Reflection(s) from Theory

Houses represent their owners and users and mean a lot to them.[27] Home is such a safe place that a person can daydream, and this is an invitation for **poetics**.[28] Houses reflect owners' economic wealth as well. Pierre Bourdieu's concept of **symbolic capital** applies to houses just as it applies to cars, jewelry, and so on. Bourdieu wrote that "with the acquisition of a reputation for competence and an image of respectability and honorability that are easily converted into political positions." Architectural representation is also a tool to achieve symbolic capital. It needs money, and in turn, it might acquire more money, power, or both.[29] Houses also represent their users' artistic tastes and their intellectuality. Theodor W. Adorno's concept of the **culture industry** (kulturindustrie) expresses an evil version of art objects' **representation of taste**.[30] The concept of the culture industry means the continuous production of new

---

27 Clare C. Marcus, *House as a Mirror of Self: Exploring the Deeper Meaning of Home.* (Nicholas-Hays, 2006).

28 Gaston Bachelard, *The Poetics of Space*, Trans. M. Jolas. (Beacon Press, 1994[1958]).

29 Pierre Bourdieu, *Distinction: A Social Critique of the Judgement of Taste*, Trans. Richard Nice. (Harvard University Press, 1984[1979]): 291.

30 Theodor W. Adorno, *Culture Industry*, 4th Impression. (Routledge, 1996[1972]).

> cultural objects to serve an industry oriented towards profit. Initially, only the upper classes could afford these objects, but later even the low-income groups could and were keen to acquire these objects. For the majority of people, it is important to faithfully follow fashion mainly because fashion expresses **economic power**. However, by the time low-income people gain access to these objects, the upper classes are able to acquire new cultural objects to express their economic power. Following the production of new cultural objects also relates to **intellectuality**.

The RONAC started to search for land and different construction possibilities before starting on the design. In addition to typical reinforced concrete building structures, there were other possibilities including container houses, and prefabricated steel and timber houses, available in North Cyprus. The main idea behind this search was to decrease the cost.

> Container houses are similar to prefabricated houses. They do not look nice, but they can be made nicer by adding some simple pergolas, and so on. The representative of the firm said that these houses are well insulated, and they have double glazing. He also told me about a little child who became healthier in one of these houses. I did not believe it because I thought any separation between these containers however small might cause disturbance. They said only 10 cm of reinforced concrete is needed for their foundations. They warned that the upper structure can separate from the foundations during an earthquake, but it will not collapse and cause fatalities. The cost of a $50\,m^2$ house was 55,000 TL (in 2016) including a kitchen, bathroom, and so on. But this excluded the cost of the foundations, the wiring, and other service connections. These were cheaper than other prefabricated houses which are almost the same price as normal (reinforced concrete) buildings. However, when everything is included, container houses can be half the price of a normal house. I also searched for timber prefabricated houses produced in Türkiye. The owner builds the foundations and then the construction team arrives to add the building within a short time. However, I was worried about working with a firm based in another country.

> I also contacted another firm which builds steel houses. One of my friends found it for me. This firm was more experienced with an artistic bent. It provided similar things and its prices fluctuated between 60,000 TL and 110,000 TL for every square meter (in 2016). I was happy with my decision for one day. But after that, when I started telling trusted friends about this

2 Context of the Project and Building and the Early Story of the House

decision, they all warned me against going in this very ambiguous direction because I had already designed a reinforced concrete house and paid for the application project and received the approval of the municipality. They said losing time by starting the whole process again, which would probably end in a similar way, would not be good decision.

> *Reflection(s) from Theory*
> The housing problem was on the agenda of many European states after World War II. However, when the role of the state changed in the production of housing, the problem of affordable housing persisted in different ways. Low-cost housing was an issue of the 1970s. The McGill University Center for Minimum Cost Housing published a special issue on **affordable housing** in Open House International in 1988. These people were seeking new ways of achieving affordable housing.[31] On the same day that the RONAC was writing these words (in 2022), CNN International broke the news: "This is a war: Californians seek affordable housing alternatives."[32] This article declared a serious increase in the number of people who do not own a house. The increase in world population, and migrations from war zones have also triggered a housing problem in the twenty-first century. The first of the 17 SDGs within the United Nations Guidebook demands "**no poverty**" and suggests "non-profit affordable housing."[33]

The RONAC decided to continue with her first project with a reinforced concrete frame system and brick walls. It must be mentioned at this point that there is generally a conservative attitude in North Cyprus to the selection of building technology. Most people prefer to have reinforced concrete frame systems with brick walls and this technology is the dominant in North Cyprus.

> *Reflection(s) from Theory*
> Responsible citizens can select sustainable building technologies. The ninth, twelfth, and thirteenth SDGs are the ones which most affect the selection of

---

31 The special issue prepared by the Mc Gill University, Department of Architecture: Center for Minimum Cost Housing, Special Issue on Affordable Housing. *Open House International* 13, no. 1, 1988.

32 David Culver and Nicole Barron, "This is a War: Californians Seek Affordable Housing Alternatives" CNN International, December 16, 2022.

33 United Nations, An Architecture Guide to the UN 17 Sustainable Development Goals.

structural material and the system. The ninth SDG demands the building of a resilient infrastructure, the promotion of inclusive and sustainable industrialization and the fostering of innovation. This objective highlights the massive amount of waste produced by the building industry and the consumption of large quantities of natural resources and energy. The use of local materials and resources, the avoidance of waste production, and the consideration of passive heating, cooling, and ventilating facilities are some of the issues to consider while selecting **sustainable building technologies**.[34] The twelfth SDG requires responsible production and consumption which demand consideration of the lifetime of the building **to avoid producing waste during the restoration and demolition** of the building and **to enable recycling**. The thirteenth SDG relates to climate action which demands the **use of renewable energies** and **designing buildings for changing climatic conditions**.

Local building industries and materials are limited in North Cyprus because most materials are imported from Türkiye. The local building materials are adobe and stone, but the building codes suggest the use of adobe and stone together with reinforced concrete elements because of the high risks of earthquakes in the region.[35] If the economy of the country is weak and there is an affordability problem, people usually prefer to select the most common building materials and technologies. The use of renewable energy can also be limited for the same reasons. There remains the option of selecting a building technology for **minimizing waste** and having **passive heating, cooling, ventilating,** and **zero carbon emissions** to achieve a better outcome.

As well as buying a low-quality (especially in terms of comfort), ready-made house, which is usually costly (because these houses are large), there are also several other possibilities for people to have a house built. Building a house might be more costly than buying a ready-made house if the site is large and valuable, and if a certain quality is to be achieved. Some people take a risk, pay the contractor, and expect them to complete the house. This decreases the cost in comparison to legal turnkey construction agreements which are less risky for clients because they pay only for the work done. Many people work with subcontractors, builders, or both and coordinate the construction of their own houses. This needs time, energy, and good contacts. Some builders, craftsmen, and subcontractors accept payment at the end of the work, but others ask for the cost of the materials earlier because

---

34 See footnote 33.
35 Yonca Hurol et al., "Building Code Challenging the Ethics behind Adobe Architecture in North Cyprus" *Science and Engineering Ethics* 21, no. 2, 2014: 381–399. Yonca Hurol, Hulya Yüceer and Hacer Basarır "Ethical Guidelines for Structural Interventions to Small Scale Historic Stone Masonry Buildings" *Science and Engineering Ethics* 21, no. 6, 2014: 1447–1468.

the cost of everything changes considerably. The RONAC started the work with a turnkey construction agreement with a construction firm. This was the same firm that built her best friend's house. The contract suggested nine months to complete the building. However, since the firm went bankrupt in the middle of construction, it took 2.5 years to complete the building. The RONAC moved into her new house in September 2019, a few months before the pandemic started.

Quality of Construction in North Cyprus

The requirements for application projects in North Cyprus are controlled by the TRNC (Turkish Republic of Northern Cyprus) Chamber of Architects, chambers of civil, mechanical, and electrical engineering, the city council (municipality), the city planning office and the electric institution. Approval for construction requires a permit from the last three of these institutions. Control is necessary during the construction process, and a professional (usually an architect or civil engineer) is responsible to the municipality for management. After the building is completed, the municipality, the city planning office, and the electric institution inspect the building, and permission to use the building is granted. However, once permission is granted, people are free to make changes to their houses.

What the RONAC experienced in regard to the quality of construction was bad. The remaining parts of this book contain sufficient information about the many problems including the application of unwanted details which were not in accordance with what was originally asked for.

The contractor of the Monarga House was not following the project instructions. This caused several mistakes such as the height of the columns and the angle of the roof, and so on. The contractor once asked the RONAC whether or not she insisted on having the booster pump as shown in the mechanical engineering project. She was told that everybody uses simple machines. The RONAC told him that she did not trust the building market in North Cyprus and preferred to trust engineers. Then the contractor had to buy the prescribed type of booster pump and ACs. However, there were many cases when the RONAC and the main controller could not change the contractor's attitude. However, the same contractor also caused some innovative and tectonic changes during construction.

The RONAC's experience was that most of the legal and verbal agreements she had made with contractors and builders were not reliable. In particular, the contractor who started the construction of the house by signing a turnkey agreement with the RONAC did not follow the instructions for most of the items in the contract. The RONAC had informed this firm that their services were no longer needed and had to get in touch with other firms/builders to complete the remaining parts of her house. During this process she understood that many contractors do not tell the truth.

Quality of Construction in North Cyprus | **55**

A minute ago, I called the electrician ... I told him that the light posts and the garage lamp existed in the project application with endorsement from all related governmental bodies. Then he told me something so that I understood why the projects in the hands of this man and me were different. He said his agreement with my contractor did not include these items. The electrical project was changed before it was given to him. I told him to prepare a proposal for me to fix these cables.

Reflection(s) from Theory

According to Albert Chan and Emmanuel Owusu's categories of corruption, changing the project agreement can be regarded as **fraud (falsification)** and **deception**.[36]

Contractors' attempts to increase their interest by making changes to projects during construction indicate that there are certain characteristics of minimizing the expenses of construction in their minds. For example, these characteristics might be related to digging less earth, decreasing the height of the ceiling, and so on. Martin Heidegger's critiques of the humans' approach to technology can explain such situations. He explained the **instrumental approach** to the world as an orientation towards the scientific approach to achieve economic benefit, and he also wrote that this happens by "enframing" objects and seeing them as a "standing reserve." **Enframing** is about defining things in specific ways (towards achieving benefit) so that people understand them as they wish. A **standing reserve** is one step further than enframing, suggesting that things in nature are there for the benefit of humans and we can go and take them whenever we wish. Jacques Ellul's theories also support Heidegger's critiques.[37] John Habraken wrote that a reflection of such an approach to economy in the building sector causes **monotonous** and **meaningless** environments. He wrote that *supplier-driven models* cause this situation rather than *user-driven models*.[38] Kenneth Frampton and Gevork

---

(Continued)

---

36 Chan and Owusu, "Corruption Forms in the Construction Industry."
37 Martin Heidegger, *"The Question Concerning Technology" In The Question Concerning Technology and Other Essays*, Trans. William Lovitt. (Garland Publishing, 1977[1954]a): 3–35. Martin Heidegger, "The Age of World Picture." In *The Question Concerning Technology and Other Essays*, Trans. William Lovitt. (Garland Publishing, 1977[1954]b): 115–154. Martin Heidegger, "Science and Reflection." In *The Question Concerning Technology and Other Essays*, Trans. William Lovitt. (Garland Publishing, 1977[1954]c): 155–182. Jacques Ellul, *The Technological Society*, Trans. J. Wilkinson. (Alfred A. Knopf and Random House, 1964). Jacques Ellul, *What I Believe*, Trans. G. W. Bromiley. (Eerdmans, 1989).
38 John Habraken, *The Structure of the Ordinary*. (MIT Press, 2000).

> **(Continued)**
>
> Hartoonian also critique the commercialization and commodification of modern architecture.[39] However, changing an agreed project is worse than Heidegger's concepts because his concepts are based on science while the corrupt situation explained above changes to the professional project.

It is difficult to find certain artisans and products in North Cyprus. The RONAC searched for a coppersmith and some parts of rain-chains and could not find any. It takes too much time to buy certain products like door handles and paint for aluminum shutters if something is chosen which is not available in North Cyprus.

After living in the house for one and a half years, the low-quality plastic rollers of the aluminum shutters had become damaged by the heavy Cyprus sun. This caused some deterioration in the shutters and the RONAC stopped using them. When she called an aluminum firm, she found that there were no high-quality shutter rollers available in North Cyprus. Actually, this was the reason which the first aluminum firm gave at the beginning. They said they could only find these small white rollers. The RONAC did not understand at that time that they would be so easily damaged. The RONAC's son sent quality rollers from abroad and the defective ones were replaced in 2021.

The RONAC's experience was that the construction work on her house was extremely slow. Every item of work was realized only after questioning it three or four times.

Although the turnkey contract asked for a penalty after nine months, the RONAC had to dismiss this contractor in the third year. She described in her diary (observation notes) a dream that she had about this problem:

> I am going to the site in my dream and seeing that the plaster team is there, the kitchen team is there, the bathroom team is there, the painters are there, and the construction finishes within two days. When I said this to the contractor he laughed and said that they would all have fights with each other … I am very worried because the construction has been stopped again.

The construction of her house stopped several times. The construction work remaining after the termination of the contact is presented in Table 2.1, which

---

39 Brian Kenneth Frampton, "Rappel a l'Ordre: The Case for the Tectonic." In *Labour, Work and Architecture*, Ed: K. Frampton. (Phaidon Press, 2002): 91–103. Brian Kenneth Frampton, *Studies in Tectonic Culture – The Poetics of Construction in Nineteenth and Twentieth Century Architecture*. (MIT Press, 1995). Gevork Hartoonian, *Ontology of Construction – On Nihilism of Technology and Theories of Modern Architecture*. (Cambridge University Press, 1994).

Quality of Construction in North Cyprus | **57**

Table 2.1 Remaining works after separation from the contractor.

Phases of building as presented in this book	Remaining construction activities
1. Application project specification	–
2. Foundations	–
3. Frame system	–
4. Walls, heat insulation, openings	–
5. Roof and pergolas	The hatch to the attic, **maintenance and protection of outdoor timber elements of the roof**
6. Plastering, painting	Painting the fireplace, the final plaster layer of the façade, the final layer of the interior walls, and the installation of outdoor pipes
7. Electrical systems	Connecting the electricity to the mains and the lamp posts at the entrance
8. Mechanical systems	The pipes and the taps in the kitchen, half of the montage work in the bathroom, **decreasing the height of manholes**, the cover and filter of the drainage well, adjusting the **solar panels and the boiler according to the mechanical engineering project instructions**
9. Ceramics	
10. Water isolation (excluding isolation in the foundations and roof)	The application of water isolation on the exterior concrete surfaces, the isolation of the sills of the triangular windows at roof
11. Windows, shutters, front door	A metal house for the pump and booster pump
12. Interior doors	The placement of glass into the top openings of the interior doors
13. Cupboards, wardrobes	
14. Fireplace, chimney	Chimney cap
15. Garden walls	**Repair of the stone wall at the front of the garden, adding a reinforced concrete tie-beam to the back of it**
16. Drainage	
17. Garage, concrete work in the garden	The concrete floor of the dog kennels, **repair of the concrete footpath to the house, repair of the steps on the west terrace, repair of the garage and garage door,** concrete elements for the garage entrance and the garden gates
18. Fences, dog kennels, garden gates	The decorative metalwork for the front door, the dog kennels, the garden fences, the garden gates
19. Pathways, roads	Filling with earth, the stabilization of some areas in the garden and the footpaths, some changes in the footpaths
20. Timber work in the garden	This item was realized by the RONAC, but it was not expected to be done before the termination of the contract

The bold items show the repair requirements or the need for change in some existing work.

2 Context of the Project and Building and the Early Story of the House

was created with the help of a document prepared by the RONAC in order to show evidence of further litigation.

There were many repairs to be done and most of these included the placement of concrete elements and the construction of specific details. The RONAC believes that she had completed one quarter of the house by making personal agreements with the builders without the use of a contractor. When construction started again the RONAC expressed how she felt:

> I started liking my house again. When it was empty for four months it looked like a graveyard to me. I lost all my interest in it, but I did not give up fighting for it. Now again sometimes I look at a photo of it on my phone and feel happy.

Reflection(s) from Theory

The concept of alienation (entfremdung) explains industrial workers' mechanistic role in production which eliminates them from reflecting their selves in their work.[40] Later, the concept of commodity fetishism replaced this concept. The fetish form represents the contrast between social/human relations and their alienated (materialized) form. Production is no longer a social relationship; instead, it is more the exchange of money and commodities. This concept also includes the mystical value of commodities and their competition with each other.[41] Thingification means forcing an intangible to become tangible, such as commodity fetishism thingifies labor relationships. Reification is treating something abstract as a concrete object.[42]

During the construction of the Monarga House, as the architect of the building, the RONAC started to feel **alienated** from her project which was an expression of herself, after the construction slowed down and stopped. This is because the change of speed of the construction reflected the mechanistic approach of the contractor towards her project or home. It can also be stated that the RONAC felt that her home was **thingified** and **reified**. But it is not possible to explain this situation with commodity fetishism, because the intangible value of the house is not lost.

---

40 Karl Marx, "Economic and Philosophical Manuscripts of 1844." In *Early Writings*, Trans. Rodney Livingstone and Gregory Benton. (Penguin Classics, 1992[1844]): 279–400. Karl Marx and Friedrich Engels, *The Holly Family or Critique of Critical Criticism*. (People's Publishing House, 2010[1845]).
41 Karl Marx, *Capital*. I, Trans. Samuel Moore and Edward Aveling volumes. (Publishing Platform, 2018[1887]).
42 Louis Althusser, *For Marx*. (Verso, 2005[1965]).

Quality of Construction in North Cyprus | **59**

I am so happy that construction flies now. I got mad till I found these builders, foremen, and workers, but it has worked. I saw so many builders, foremen, and workers each day in order to find a good one and to reach an agreement. And finally, I am happy with them.

The RONAC's experience of the construction work was that it was terribly unprofessional resulting in considerable damage to the ceramics in and around the building and the plants in the garden. She wrote that:

I cannot stand the idea of having mess at my place and every day it is getting worse. Workers do not collect rubbish somewhere to throw it away later. They just throw it anywhere. They also do not care about the trees. One of them cut a branch, after I had asked him to, but he cut it in a very bad way. Later they started nailing things to the trees. I asked them to be removed. Once they lit a fire under one of my trees ... I argued with one of the workers because of the mess created with gypsum. I was worried about having gypsum around my trees because I want a green garden. I asked for it not to be done again. I hope they will not.

I had to clean my site. Yesterday I collected two big plastic bags full of rubbish from the vicinity of the site. So, I put them on the pavement to be taken away by the municipality rubbish truck.

The relationship with foremen and workers could also be unpleasant.

This is the man who was using my house as a toilet. Then he became an Islamic fundamentalist for a while and now he has changed back again. In any case he likes worrying me.

... This type of messy work is never acceptable. We argued with one of the painters. He said I had accepted their work and they have already been paid. I told him to go away then. He also interrupted me. Then I interrupted him. Then I shouted at him. They cleaned up their messy work and left. However, ... somebody else did most of the work. Not them.

... the ceramic placed in the square area beside the window was almost colorless. The worker found the most colorless ceramic and put it there. I told him clearly that I wanted something colorful there. But he is a strong character, and he wants to decide. But this is not acceptable. He got angry with me when I told him this because he was planning to leave with the work completed. He thought his work was ready for the electrician. But because of this change it is not.

... I have a feeling that these people behave like this because I am a single woman alone. They think I cannot argue with them.

> *Reflection(s) from Theory*
>
> **The percentage of women in construction** in the United States is 10.9% due to several reasons including safety.[43] It might be worse in North Cyprus. The RONAC also had some difficulties during the construction of her house. Although there were a few aggressive builders, foremen, and workers, her problems mainly originated from the bankruptcy of the contractor. She also found that people (and some people even in her family) generally believed that her involvement in construction activity could not produce good results.
>
> The belligerent attitudes of builders, foremen, and workers might be related to the **general belief that the majority of innovations are carried out at the project level by professionals.**[44] This expectation can reduce the innovative attitudes of builders, foremen, workers and prevent them from being reflected into their work. Although it is also thought to be good to open up the innovation process to all workers,[45] it is clear that not much of a contribution is expected from builders, foremen, and workers. However, contrary to the above sources, the preliminary design of the Monarga House did not contain any innovations and **contractors, builders, foremen** had considerable **innovative attitudes** during its construction.

Horror Stories about Building a House

One of the RONAC's colleagues told her that many construction companies in North Cyprus leave their clients towards the end of construction. The RONAC thinks that the contractor for her house also used this tactic. He first gave her a low price in Turkish lira. He then increased the price when the architect of the application project said that he had not considered the material and system suggestions given by the RONAC. He also asked for 60,000 TL in cash to begin the construction (in 2017), and said he wanted to subtract this money from the amount agreed from the later phases of the construction process. The RONAC asked him to subtract it from the early payments which were defined in the agreement, but

---

43 Buildern, 7 Major Concerns Women in Construction May Have, 2021, accessed February 18, 2025. https://buildern.com/resources/blog/7-major-concerns-women-in-construction-may-have/.

44 Xiaolong Xue et al., "Collaborative Innovation in Construction Project: A Social Network Perspective" *KSCE Journal of Civil Engineering* 22, 2018: 417–427.

45 Henri Simula and Tuomas Ahola "A Network Perspective on Idea and Innovation Crowdsourcing in Industrial Firms" *Industrial Marketing Management* 43, no. 3, 2014: 400–408.

she did not check this. That is when it becomes easy for a contractor to abandon the work towards the end of construction because they are no longer interested. Actually, such contractors force clients to dismiss them by slowing down the work.

> *Reflection(s) from Theory*
> According to Albert Chan and Emmanuel Owusu's categories of corruption, slowing down the work as planned from the beginning to force the client to dismiss them towards the end of the construction can be regarded as **coercion.**[46]

The RONAC heard some terrible stories about contractors during this process. A notary told her that most of the court cases are against contractors and an old student of the RONAC told her that the laws of North Cyprus protect contractors. If a person makes a new agreement with a different contractor after dismissing the first one, the work to be done should be itemized, because this can be helpful if the court sends experts. Otherwise, experts could make their own categorization, and this might place the owner in trouble. There should be a controller report for each payment and receipts should be kept even if there is an official agreement between these people. Otherwise, the first contractor might demand money claiming that they have completed the construction although they have not.

Some newspaper stories tell of British people who lost all their retirement savings while trying to buy a house in North Cyprus. They thought that North Cyprus had similar laws to English ones which protect people. There was also news about some apartments which were paid for but not finished for many years. One of the workers of a firm told the RONAC that they had given all their money to a steel construction firm and the firm became bankrupt after building the skeleton of their house. It left the country two years ago. This means that after being exposed to nature for two years, that structure became useless. Horrified by such stories, the RONAC wrote in her diary:

> I started to think that somebody should research home building problems, because many people have had very serious problems. Many people have lost money while trying to build their homes. Many people in the sector have been bankrupted because of corrupt people. Two carpenters closed their businesses because of my contractor as they were not paid. Many people have their houses built by small firms but not by contractors. Contractors ask for very high interest for sure. The economy is not reliable here.

---

46 Chan and Owusu, "Corruption Forms in the Construction Industry."

Yesterday we were talking to one of my students and he told me that in Marshal Berman's book of *All that Solid Melts into Air* he wrote that secularization happens properly in western countries and business runs properly, but for most of the underdeveloped countries, where secularity is mixed with tradition, there is big trouble going on in business. Tradition demands trust and kin relationships, while secularity requires objective bureaucratic business. When these are mixed, there is trouble. I organized everything according to modern business rules, but my contractor might have kinship relations to protect himself.

Reflection(s) from Theory

Secularity can be defined as being related to the physical (material) world and not to the spiritual (religious) world. Secularity changed the authority of religion and fueled science, technology, and the productivity of humans. This is the main difference between the **two paradigms: the traditional and the modern.** This paradigm difference also changed the speed, the type of knowledge required (standard scientific knowledge), and the type of education (mass education instead of the master–apprentice system) within the new era of large-scale production. Many professionals, including architects and engineers, are needed to meet such a large demand for production. Standardization of knowledge enables and eases professional communication, control and management of production procedures. Traditional building production was slower and based on practical knowledge. There was the authority of master builders and experienced older people. As the roles of people were defined within their religion, the relationships between them were not objective, as they are now within modern professions. Marshall Berman explains modernity with the help of many scenes from early modern social life including professional relationships.[47]

After terminating the contract, the RONAC also experienced this problem with the builders and foremen. The RONAC preferred to have written agreements, while the majority of builders and foremen were worried about signing such agreements. The RONAC explained to them that such agreements protect both sides; however, they were still worried.

Builders and workers are stressed because they might not be paid. Some of them think this is normal. They say that they build and if the client likes the outcome, s/he pays.

---

47 Marshall Berman, *All that is Solid Melts into Air: The Experience of Modernity*. (Penguin Books, 1988[1971–1981]).

> I have a feeling that carpenters are in trouble here and this makes them nervous.

When the RONAC dismissed the contractor, she herself was subjected to this problem.

> later I started to worry that the people who were not paid by the contractor will ask for money from me.... Everything was arranged but I ended up with a shock, because they asked me to pay the total price twice including the money I payed to the contractor. The people I had communicated with were not there.

Disasters, Climate, and Building Culture in North Cyprus

North Cyprus is in a high-risk earthquake zone. The RONAC, who has researched earthquakes, preferred to buy land with a rocky soil type to minimize earthquake effects. There is also a danger of wildfires because of climate change. Wildfires are dangerous especially for buildings in rural areas. The RONAC's house stands alone in a field. During June 2022 there was a fire in the mountains of North Cyprus and the fire reached the village adjacent to Monarga. The RONAC was able to see the fire from her windows.

Reflection(s) from Theory

The Mediterranean Sea and its surroundings are a **high-risk earthquake zone**.[48] There are also **climate related disasters** which have tripled since 2006. Sea levels and air and water temperatures have risen. Cyclones, tsunamis, wildfires, droughts, and floods are more frequent and severe than they were before. The main reason behind these climate related disasters is the effect of carbon gas emissions which build up in the atmosphere. The IPCC (Intergovernmental Panel on Climate Change) 2022 Report studied all factors which cause increases in carbon gas emissions in the atmosphere and invites action.[49] It is also felt in Cyprus that there are more **floods** and **wildfires** than before.

---

48 Yonca Hurol, *The Tectonics of Structural Systems – An Architectural Approach*. (Routledge, 2016).
49 IPCC In *Climate Change 2022: Mitigation of Climate Change. Contribution of Working Group III to the Sixth Assessment Report of the Intergovernmental Panel on Climate Change*, Eds: P. R. Shukla et al., (Cambridge University Press, 2022).

2 Context of the Project and Building and the Early Story of the House

Many people think and say that the quality of buildings in North Cyprus with respect to climate is very low. Most houses have reinforced concrete frame structures with brick walls, windows and doors and flat roofs without any heat insulation. These houses are like ovens, especially on summer nights. Sometimes the construction quality is so low that one can feel the winter wind inside the house. For this reason, people use too much electricity during winter and summer and pay high electricity bills.

> *Reflection(s) from Theory*
>
> The 2018 International Energy Conservation Code (IECC) describes how different parts of buildings in the United States should be insulated.[50] TS825 2008 is about the thermal insulation requirements of buildings in Türkiye.[51] The aim of this standard is to reduce energy consumption in buildings. Although North Cyprus has many parallels with Türkiye, the quality of buildings in terms of thermal insulation is extremely low. Khaled Shanablih wrote a master's thesis on the **energy efficiency of houses** in North Cyprus in 2012.[52] He sent a questionnaire to 231 individuals and found out that 94% of houses have no wall insulation and 89% have no roof insulation. He also found that energy loss through the walls is 98% and through the roofs is 63% greater than European recommended values. Shanablih suggested having "3 to 7 cm polystyrene or glass wool insulation for terracotta brick walls and concrete roofs, respectively, and double-glazed windows will be enough to reduce the heat losses to European standards."
>
> On the other hand, houses with **passive heating and cooling** are also possible. Careful consideration of orientation, heat insulation, passive house window design with airtightness, ventilation with heat recovery and the avoidance of thermal bridges are the basic ways of achieving passive heating and cooling in houses in Cyprus.[53]

When the RONAC was looking for a ready-made house, she described the main characteristics of the house that she wished to buy: a small house with a pitched

---

50 International Energy Conservation Code, 2018, accessed February 16, 2025. https://codes.iccsafe.org/content/IECC2018P5/chapter-4-re-residential-energy-efficiency.

51 TS825, Thermal Insulation Requirements for Buildings 2008, accessed February 16, 2025. https://sayfam.btu.edu.tr/upload/dosyalar/1458664642TS-825_Standard.pdf.

52 Khaled A. Shanablih, Determination of Energy Efficient Construction Components for Houses in Northern Cyprus (Master's Thesis, Near East University, 2012), accessed December 21, 2022. http://docs.neu.edu.tr/library/6296379156.pdf.

53 Passive House and NZEB Design, accessed December 21, 2022. https://passivehouse.cy/.

roof to protect the house from heat, set in a large garden. Her previous house had a flat roof, and she wanted a single-storey house to avoid stairs. She also preferred to have places in the garden which were distant from neighbors' houses to avoid disturbance and be able to keep her dogs safe. She could not find a house like this.

When she decided to buy land and design a house, she became more demanding about climatic comfort. She wanted some land with trees to enjoy the shade. She wanted to design a comfortable house with a good orientation, a pitched roof, high ceilings like traditional Cypriot houses, and heat insulation on the walls, roof, and windows and doors. Although she wanted to have a small building hidden behind trees, she had to make the ceilings high to achieve better comfort in her old age.

> *Reflection(s) from Theory*
>
> There can be various levels of small buildings hidden in nature. Gaston Bachelard describes an extreme condition for being hidden in nature under the heading of **"house and universe."** He says that if there is danger in the natural environment and if people find a small shed hidden in nature, this can cause poetic feelings.[54] However, the RONAC's Monarga House cannot cause such feelings because it is not that small. The environment around the house is also not wild. The house is in a village. Comfort conditions were more important for the RONAC than the poetics of the house.

I also started telling my mother that she can come and stay in this house without having temperature problems and without any need to use air conditioners.

Physical and Social Environment of the Building

Before buying the land in Monarga, the RONAC refused plots of land in this village. This was because she had problems with her neighbors there. Otherwise, she liked the physical environment of Monarga and built bonds with it during the years she lived in the Maronite's house. She was very happy with the sea view, the scenery, and plants, trees, and animals ... However, she bought some land very close to her previous house and with the same neighbors.

---

54 Bachelard, *The Poetics of Space*: 38–73.

> **Reflection(s) from Theory**
>
> Christopher Alexander wrote that "when you build a thing you cannot merely build that thing in isolation, but must also repair the world around it, and within it, so that the larger world at that one place becomes more coherent, and **more whole**; and the thing which you make takes its place **in the web of nature**, as you make it."[55] Therefore, the RONAC did not have a good beginning because of her negative feelings towards her neighbors. Although she liked the physical environment, she did not feel at ease with the place.

Most of the buildings around the RONAC's site are simple old houses abandoned by Greek Cypriots. These houses are on large plots of approximately 900 m². The houses are small and most have only one floor. Many of them have additional coach houses in the back garden. Most people living around the RONAC's site are originally from Monarga. Their houses and farms were burnt down in 1963 and they returned to different houses in 1974.[56] There are five houses in the same street as the RONAC's house. Men who live in these houses are either carpenters, taxi drivers, builders, or electricians while the women are housewives. Most of the men are hunters. They are part of a close network with each other.

> **Reflection(s) from Theory**
>
> Monarga is a heterogenous village when considering its population. However, each part of it is heterogenous in a different way. The part of Monarga where the RONAC located her house, is one of the most heterogeneous parts of the village, considering its population. There are people from Türkiye as well as former residents of Monarga, and other Turkish Cypriots. There is heterogeneity due to income levels as well as their value systems. Some are people with contemporary values and others are villagers trying to live in the same way as their fathers and grandfathers. Amos Rapoport discussed the advantages of homogeneity and heterogeneity. According to him, some commonality is needed for interaction between people. The identity of the environment also becomes clearer in **homogeneous environments**. However, homogeneous environments can be small. **Heterogeneous environments** need neutral spaces (such as parks and grocery stores, etc.) for different people to interact.[57] There are no **neutral spaces** in Monarga. There is one village locale, but it cannot create interaction between everybody.

---

55 Christopher Alexander, *A Pattern Language – Towns, Buildings, Construction*. (Oxford University Press, 1977).
56 Hurol and Farivarsadri, "Reading Trails and Inscriptions."
57 Amos Rapoport "Neighbourhood Heterogeneity or Homogeneity – The Field of Man-Environment Studies" *Architecture and Behaviour* 1, 1980: 65–77.

The streets are not busy in this part of Monarga. There are regularly placed houses across the road from the RONAC's land. However, her side of the road has only two buildings which are set apart from each other. Later, the RONAC's son told her that this was the site where a Turkish airplane crashed during the 1974 war. The plane was hit by anti-aircraft fire from guns placed on the roofs of these old houses. The RONAC remembers that the old lady living next door to her previous house had shown her the propeller of this plane.

There are two telecommunication base stations on the street parallel to the RONAC's street. Actually, these base stations were adjacent to her previous house. Although there is contradictory information about their danger to health, the RONAC worries about them. She researched this subject some years ago and found that the base stations in North Cyprus were not very powerful. However, later on, the army loaded these base stations with extra equipment, and they became dangerous. The new site is 180 m away from these base stations which is less than the defined safe distance of 250 m.

There are interesting and attractive trees around the village, such as large pines, carobs, palms, almonds, pomegranates, lemons, and more. There are also bougainvillea (*cemile* in Turkish), China roses, and similar small plants. Other parts of the village also contain rubber and fire trees.

About the Site

The site is 37–38 m above sea level according to the calculations of a friend who can get satellite information. It slopes towards the south-east. Since the site is very rocky, digging the foundations would be costly. The site has a good view of the sea to the east and south. However, if the three plots adjacent to this land are built on, the sea view might be partially obscured.

Reflection(s) from Theory

The concepts of *topos* (the site), *typos* (the meaning), and *tecton* (carpenter) are studied by Kenneth Frampton. His concept of *topos* has strong connections with culture. Frampton wrote that **topography, climate, and culture** effect the tectonics of buildings.[58] Yi-Fu Tuan suggested the concept of **topophilia**; *topos* meaning the place and *philia* meaning love to express the connections between the material world and humans.[59]

---

58 Brian Kenneth Frampton, *Studies in Tectonic Culture – The Poetics of Construction in Nineteenth and Twentieth Century Architecture*. (MIT Press, 1995).
59 Yi-Fu Tuan, *Topophilia: A Study of Environmental Perceptions, Attitudes and Values*. (Columbia University Press, 1990).

2 Context of the Project and Building and the Early Story of the House

There were some small pigeon and chicken coops, and also some rubbish (old cars, bottles, stone pieces, etc.) on the land. The RONAC thought these could easily be taken away by their owners. She asked them twice to collect them from the land before she started building her house. She waited for two months and asked her lawyer about what to do. He suggested she should go to the police. After a policeman came and took notes, these neighbors eventually moved their coops from the site.

There were some quite large pepper trees at the front of the site. There was also a nice, shady, hidden corner under one of these trees. Since there were coops on the site for quite a long time, there was an interesting pathway between them. Later, the RONAC decided to use this pathway as the main entrance route to her house. This is because she decided to put the house behind the trees and not to cut them down. The RONAC also liked the idea of hiding behind these trees and use them as shields to protect herself from the dangerous effects of the two base stations. She thought simultaneously that this might cause some security problems if she was too concealed.

Reflection(s) from Theory
Consideration of context is essential for architecture, and it is also generally accepted that contextuality is strongly related to **architectural ethics**.[60]

The RONAC found out that the placements of trees were not correctly shown on the site plan which was drawn with the help of satellite information. She thought that this had possibly happened because the trees are not straight. They start at one point and bend in other directions because of the wind. So, they look as if they are at other points, but their trunks sit in different places. The RONAC kept most of those trees.

Reflection(s) from Theory
The philosopher Martin Heidegger's writings contain his thoughts about architecture which also reflect his critiques about the **impact of architecture on nature**. His words about a bridge at the fringes of a settlement can be seen as an expression of this critique. He also relates ethics and a consideration of context in a holistic way and asks for the bringing together of the **fourfold aspects of earth, sky, divinities, and mortals** which enable one to participate

---

60 Thomas Fisher, *The Architecture of Ethics*. (Routledge, 2018).

> in this world. Earth symbolizes the material world and life's suffering. Sky embodies human thought, encompassing analytical, artistic thinking, and contemplation. Divinities signify ancestral signs and cultural heritage. Mortals remind humans of their own mortality. Heidegger also wrote that architects should design buildings *as they were*. He meant that new buildings which will replace collapsed buildings due to wars, should be designed to resemble those damaged buildings.[61]

Conclusion

Concerns about the political economic context, the context of building design and construction, the quality of construction, horror stories, the context of disasters, climate and building culture, the physical and social context of the building and the site can be retouched as a conclusion with the help of reflections from theory.

> *Reflections from theory about the political economy* of North Cyprus contained **political** keywords such as conflict zone, transitional justice activities, multiculturality, symbolic compensation platforms, nationalism, traces of war, and nomadic wartime equipment.
>
> There were some **philosophical** keywords: uncanny geography, identity thinking, territorialization, deterritorialization, and reterritorialization.
>
> There were some **construction** keywords: corruption in the construction sector, a supplier refusing to deal with anyone other than their main point of contact, and the avoidance of all types of corruption within the construction industry.
>
> There were also two **architectural** keywords: problems of place attachment, struggling for a lost home.
>
> There were some **architectural and political** keywords: alternative cultural heritage, resettlement.
>
> There was one key word from the **literature**: experiencing the same haptic affects.

> *Reflections from theory about the context of building design and construction* contained many keywords about **sustainability**: no poverty, sustainable

(Continued)

---

61 Martin Heidegger, "Building Dwelling Thinking." In *Poetry, Language, Thought.* (Harper and Row Publishers, 2001[1954]): 147–148.

(Continued)

building technologies, minimizing waste, avoiding waste production during restoration or demolition, enabling recycling, the use of renewable energies, designing buildings for changing climatic conditions, local building industries and materials, passive heating, cooling, and ventilating, zero carbon emissions.

There were some **philosophical** keywords: symbolic capital, the culture industry.

There were some philosophical, architectural, and **aesthetical** keywords: poetics, expression of economic power.

There were some **architectural** keywords: representation of taste, affordable housing.

Reflections from theory about the quality of construction contained keywords about **construction**: fraud (falsification), deception, the percentage of women in construction, a belief that the most innovations are carried out at the project level and by professionals, innovative ideas of the contractor, builders, and foremen.

There were some keywords from **philosophy**: the instrumental approach, enframing, standing reserve, speed, being alienated, thingified, reified.

There were some key words from **architecture**: monotony, meaninglessness.

Reflections from theory about horror stories contained one keyword from **construction**: coercion.

There was one key word from **philosophy**: two paradigms of traditional and modern.

Reflections from theory about disasters, climate, and building culture contained keywords combining **architecture and structural engineering**: high-risk earthquake zones, climate related disasters, floods, wildfires.

There were some **architectural** keywords: energy efficiency of houses, passive heating and cooling.

There was one **philosophical** keyword: house and universe.

Reflections from theory about the physical and social environment of the building contained only **architectural** keywords: being more whole (in the web of nature), homogenous and heterogenous environments, neutral spaces.

> *Reflections from theory about the site* contained some **architectural** keywords: topography, climate and culture, topophilia, consideration of context, impact of architecture on nature.
>
> There was one **philosophical** keyword: fourfold aspects of earth, sky, divinities, mortals.
>
> There was one **architectural and philosophical** keyword: architectural ethics.

The RONAC experienced concerns in her roles as an owner, architect, and controller. A search of her diary for the word "worry" reveals numerous entries, most of which reflect her concerns as the controller. Some of these are discussed in later chapters. Meanwhile, her worries as an owner are often expressed through the word "dream."

> One night ... I woke up in my dream and went into the garden to see why the dogs were barking. I found that my garden was full of cars. People had turned my garden into a parking place. I was very upset. A nightmare.

> A few nights later, the dogs were barking again. I did not wake up and continued to sleep. But in the middle of my sleep, I heard a big noise. I jumped. I thought this might be from the leaning tree in the garden. I thought that maybe it had fallen down ... Actually, the noise and barking dogs were parts of a dream ... Later, the dogs really barked, and I went out. I wanted to check the tree too. The tree was still alright. Then I started worrying about the chimney and the trusses of the garage. Later I checked them too and understood that the noise was in my dream.

3

Preliminary Design of the House and Tectonic Affects due to Changes in the Application Project and the Tendering Process

The Monarga House is a small one-story house which has a reinforced concrete frame system, eco-brick walls, aluminum window frames and shutters, and a pitched roof with pergolas. Most of its façades are plastered using a local special type of plaster called leaf plaster. The house is located towards the back of a 900 m^2 plot with many trees. Most parts of the garden are naturally maintained by using stabilized earth and gravel, avoiding the use of concrete surfaces and protecting the existing trees. The owner of the house is also the researcher who wrote this book, a neighbor of the construction site of this new house, the architect, and one of the controllers (the RONAC, Researcher, Owner, Neighbor, Architect, Controller). This book does not present this house as a successful example of architecture. In fact, it contains information on the many problems in the design and building processes. The building was not designed with any innovative intent. However, starting from the application phase there were a number of innovative attitudes.

The RONAC's diary (observation notes) was coded using the concepts of *image, picture, feel, freedom, style, color, four directions, tectonics, mother, domestic,* and *minor* in order to write about the preliminary design phase in this chapter. Most changes during the application stage and the tendering process were extracted from the RONAC's diary with the help of the coding concepts of *change, idea, problem, contribution,* and *trouble.* Changes causing tectonic affects, innovative attitudes, or both were analyzed separately. Then the feelings of the RONAC were also inserted into the text together with reflections from theory. The first half of Chapter 3, which is about the preliminary design phase, has four layers (the drawings and photos, information about the preliminary design, the feelings of the RONAC, and theoretical reflections) and the second half of which is about the application phase and the tendering process, contains six layers (the drawings and explanations, changes, tectonic affects, innovative attitudes, the feelings of the RONAC, and reflections from theory).

Tectonics as a Process in Architecture, First Edition. Yonca Hurol.
© 2025 John Wiley & Sons, Inc. Published 2025 by John Wiley & Sons, Inc.

Preliminary Design Ideas and Main Tectonic Decisions

Before designing this house, the RONAC tried to buy a ready-made house. This was her first choice because of the complicated problems of having a house built in North Cyprus. Later, her preferences within this process had some effects on the design ideas and decisions of the Monarga House. She was looking for houses on a firm soil type, because Cyprus is in a high-risk earthquake zone. As one of the RONAC's academic research interests is structural systems and earthquakes, she wanted to use her knowledge for her own house. Because of this, she avoided certain districts with weak soil.

One of her other preferences was not to buy a house with a luxury image. She was looking for a small, basic house preferably set in a large green garden to keep her four dogs fit. Another of her preferences was to have a one-story house or at least a house with accessible stairs. This is because she was concerned about her quality of life in old age. She wrote the following statement about this in her diary:

> I also remembered my mother's problem with the stairs. When she became old, it became difficult for her to climb four floors and she had to stay at home. Her health deteriorated, and her social skills were also affected badly mainly because of this.

The RONAC's fourth preference was to have a pitched roof and pergolas covered with red tiles. She liked houses with red pitched roofs, and she knew that flat roofs make the climatic conditions in buildings very difficult, not only during Cyprus summers but also winters. There were houses with pitched roofs, but most of the houses for sale, were large expensive buildings with extremely awkward stairs and had small gardens which meant close proximity to the neighbors' houses. Therefore, she decided to buy land and design a house for herself.

She started looking for a large area of green land with firm soil, not far from Famagusta or Iskele (Trikomo) to be close to her workplace. She was also looking for land away from the main roads to avoid any danger of traffic for her dogs and cats. Since she had rented in Monarga since 2005 and she did not like her neighbors, she preferred not to buy land in Monarga. However, Monarga has firm soil and contained many large green plots of land for sale. Finally, in 2015 she bought a large plot of green land very close to her previous house which also has good sea and mountain views. It was possible to see this land from the windows of the house that she was living in at that time. Living adjacent to her construction site was to be a big advantage in the future while controlling the construction of her new house.

The main design ideas of the Monarga House which was designed by the RONAC, were not elaborate, These include a living area with wall openings in four directions (east, west, south, and north), a central fireplace, a study area, and a small dining space within the living area, and fenestration to make this living area completely accessible to the garden; a climatically comfortable house; a simple traditional image; full accessibility without any steps; small and cheap to run; concealed at the back of the green garden; appropriate for pets, concern about the view because of the possibility of losing it in the future; the importance of security, selecting materials, products, colors in an indulgent way; and some tectonic details. Figure 3.1 demonstrates the preliminary architectural project which was designed with these ideas in mind and two photos of the completed house looking at it from the outside. Figures 3.2 and 3.3 present photos of the house's exterior and interior.

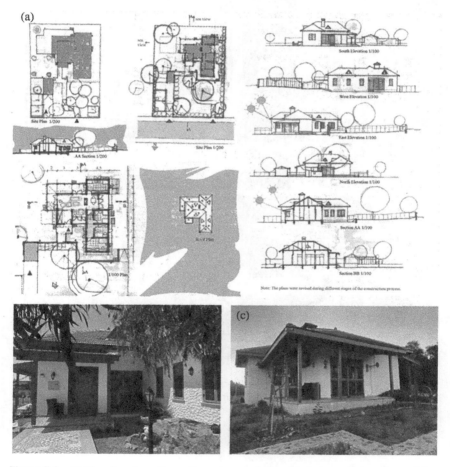

Figure 3.1 (a) The site plan, site section, plan, roof plan, sections, and elevations of the Monarga House. (b) View of the entrance of the house. (c) View of the house from the back garden.

76 | *3 Preliminary Design of the House and Tectonic Affects due to Changes*

Figure 3.2 (a) Looking towards the west terrace from the living area. (b) Looking inside from the south terrace. (c) The corridor leading to the bedrooms. (d) Looking inside from the north side. (f) Looking towards the east and south terraces from the living area.

Figure 3.3 (a) The fireplace in the living area. (b) A footpath in the garden. (c) From the living area. (d) Looking towards the entrance of the house from the garden.

A Living Area Opening in Four Directions

The strongest initial idea of the project was to have a living area with wall openings in four directions. This area also included a study and a library which would actually be one of the main functions of that place, and a dining area. The RONAC's first design in this direction was symmetrical and it looked like a temple. She did not like that image and tried to achieve a more contemporary one without changing the main character of the living area.

The room being open to the four cardinal points represents the world view of the RONAC in respect of not discriminating against people and respecting human rights as well as the rights of other beings; being open to them. The level of openness of the main living area was another concern. The living area was intended to be a part of the garden in the summer when the glass patio doors were opened.

> I designed my house facing four directions with large glass patio doors. When these panels are opened the place will be like an open space.

At the beginning, some tectonic details also contributed to this main design idea. These included large accordion windows to be fully opened, entirely openable shutters, a smooth floor between the inside and outside, the use of the same floor materials inside and outside, and light garden furniture inside the living area. The effect of designing for one's self can also be reflected strongly in these decisions. However, some of these ideas were changed very early on. It should be noted here that the preliminary design in Figure 3.1 contains a later solution for the windows and shutters.

This decision to have a living area as a part of the garden also was related to to the climatic comfort of the building.

> I wanted to open those façades totally in summer (for cross ventilation) and to close them with shutters as needed in winter and for security.

The fire was intended to heat the whole house when lit, so the fireplace was not connected to the outside walls to avoid heat loss.

> I also want to have a large two-sided fireplace … in a central place … in the living area. This will be the *hearth* of the house.

Reflection(s) from Theory

Prairie architecture which appeared during the late nineteenth century, was inspired by nature (as Frank Lloyd Wright said "married to the ground"), craftsmanship, and simplicity. These buildings have strong massing, a central fireplace, brick or stucco exteriors, open and asymmetric floor plans, connected indoor and outdoor spaces, the use of wood in indoor spaces, and the exploration of motifs (e.g., in plaster, wood …).[1] The Monarga House also has a central fireplace, an open and asymmetric floor plan, connected indoor and outdoor spaces and an exploration of the leaf plaster motif.

Comfort

One of the main reasons for having a climatically comfortable house was to have high ceilings and a cool roof, which is a ventilated pitched roof over a reinforced concrete ceiling that is also covered with heat insulation. The house was also to

---

1 Rachel Brown, "Prairie Style Architecture: The Iconic American Style," 2022, accessed February 16, 2025. https://www.homedit.com/house-styles/prairie-style-architecture/.

Preliminary Design Ideas and Main Tectonic Decisions | **79**

have large pergolas, especially one on the south side of the house to stop sunlight from coming inside during summer and to let it in during winter. The change in the angle of the sun in summer and winter and its effect on the interior space can be seen in one section and one elevation of the preliminary design project in Figure 3.1.

> I decided to have a ceiling height 3.5 meters in order to support the concept of having an open space. The height of the pitched roof is another issue. It must be possible to enter the attic because there will be a need for a water tank in the attic which will also be used when there is power cut. All these affected the image of the house. Although I wished to have a small building lost in the trees, I had to make it high.

The RONAC wanted to have a comfortable house with good orientation and proper heat insulation. Apart from those windows in the living area, the windows were small and minimal to the west and north. This resulted the back façade facing west. Although the prevailing wind direction is west, sun coming from west is dangerous in summer; and the north wind is dangerous in winter. She also preferred to have heat insulation on the walls and decided also to have window and door mullions with heat insulation. All openings were planned to be double glazed.

> Although I had doubts about having heat insulation, finally I decided to have it whatever its price as money and image.

These decisions gave the house an almost traditional image. However, according to the RONAC comfort has multidimensions. In her diary, she expressed how her completed house was comfortable as follows:

> I was telling one of my friends that my life is easier in this new house. She asked how? I started thinking about this. I think the first thing is about having better conditions for animals. Then having many wardrobes and cupboards and a tidy environment is lovely. Then it is easier not to have stairs. I was going up and down many times in the other house. This solved my knee problems within two years. I can see everything from my windows. The garage provides many facilities. The water system is much better. There is high water pressure, and I will wash my car today. It is also much easier to clean my dog kennels because of this. The doors of the kennels can be opened and closed easily. I use less AC, because the house temperature is acceptable due to its good insulation and orientation and there is very nice breeze everywhere during summers. So, it is also more economical.

Currently the house temperature is almost comfortable even during the worst days of winter and summer. Since the living area has cross ventilation in both directions, there is no need to use electricity during the day. On summer nights, if the windows are closed, it becomes necessary to use the ACs for a short time. The electricity bill for the house is really low when compared to the bills of others.

> *Reflection(s) from Theory*
> The thirteenth Sustainable Development Goal (SDG) in the 2018 UN Guide for Architecture, suggests climate action which demands the use of renewable energies and the design of buildings for changing climatic conditions, having **passive heating, cooling, ventilating,** and zero carbon emissions to achieve a better outcome.[2] The RONAC's intention was not to design a sustainable house, mainly because of economic problems. However, one of her aims was to have some passive heating, cooling, ventilating facilities.

Use of Leaf Plaster

There is a small Greek house in Monarga which inspired the RONAC during the design phase of her house. That house has a large garden with many trees, a pitched roof with red tiles, and dark green shutters. It is white, and it has leaf plaster on its walls. Leaf plaster was used in many houses in Iskele/Trikomo and generally in Cyprus. The RONAC wanted to use this old technique of plastering for her house too.

> *Reflection(s) from Theory*
> Gevork Hartoonian used the concept of *métier* for the reinterpretation of old building techniques and exemplified it through Frank Lloyd Wright's architecture. He reinterpreted several old building techniques in his architecture. The use of stone in Fallingwater is an example of this. The RONAC also wanted to use the old leaf plastering technique.[3]

Accessibility

One of the very early design decisions was having only one level. The RONAC never forgot the difficulties her mother had while spending her old age in an apartment on the fourth floor without an elevator.

---

2 United Nations, *An Architecture Guide to the UN 17 Sustainable Development Goals*, 2018, accessed February 16, 2025. https://www.uia-architectes.org/wp-content/uploads/2022/03/sdg_commission_un17_guidebook.pdf.
3 Gevork Hartoonian, *Ontology of Construction – On Nihilism of Technology and Theories of Modern Architecture*. (Cambridge University Press, 1994).

Although there is a slope on the site towards the south-east direction, I decided to have no stairs inside the house, as I was thinking about my old age, since I was 56 at that time. While designing the site this idea went further, and I designed the house in such a way that I can go in and out without any stairs or steps. There are some stairs between the terraces and the garden, but they might not be used. All the differences in level are covered by ramps.

Finally, there are no stairs or steps between the garden gates, garage, and the entrance to the house.

Reflection(s) from Theory

The tenth of the 17 SDGs demands the reduction of inequalities by considering human rights, including the rights of disabled people. This can be reflected in architecture by providing **accessibility** for disabled people and universal design to consider the needs for everybody.[4] The RONAC was firm about designing an accessible house for herself.

Small and Economic

Since this house was going to be built with the RONAC's retirement savings, it was important and inevitable to have a small, easy to run house. Otherwise, there was a danger of not being able to complete it. Another reason for having a small house was to make it easy to clean. The image of the house was also small in her mind. Only the ceilings were high to provide comfort in hot weather.

Reflection(s) from Theory

The eleventh of the 17 SDGs demands sustainable cities and communities which require consideration of **affordability**, climate change, and **elimination of loss of vegetation and biodiversity**.[5]

A small house in a natural environment is also a reminder of Gaston Bachelard's "**house and universe**" which explains the poetic feelings of taking refuge in a small mountain hut during a violent storm.[6]

(Continued)

---

4 United Nations, *An Architecture Guide to the UN 17 Sustainable Development Goals.*
5 See footnote 4.
6 Gaston Bachelard, *The Poetics of Space,* Trans. M. Jolas. (Beacon Press, 1994 [1958]).

> *(Continued)*
>
> The RONAC was decisive about designing an affordable house which also eliminates the loss of vegetation and biodiversity. Although she designed a small house in nature, this is not comparable with the "house and universe" of Bachelard.

Hidden at the Back

There are pepper trees on the front part of the site. Many Cypriots think that pepper trees have no value. The RONAC disagreed with that. She preferred to protect these trees and placed her house towards the back of the site. This also matched her idea of rejecting ostentation and having a small, concealed house reached via a footpath.

> I also wished to eliminate any sort of ostentation including economic prestige, power prestige, and also architectural prestige which might end up with a sort of architectural gymnastics of unnecessary elements or forms. I always disliked being identifiable. Also, I have to design something inexpensive especially after buying this large plot of land which was cheap in itself but still expensive for me.

Keeping the garden as natural as possible was part of this idea. Minimizing the use of concrete in the garden was also important.

> I wished to have a natural footpath and a drive on the site. So, I wanted to follow the suggestion of the architect of the application project to have a stabilized drive using yellow Cyprus stone dust.[7] I also wanted to have stone pieces to form the footpaths on the site. There will be few footpaths. I prefer to walk on earth mostly. Cover is needed only for wet weather to avoid walking in mud.

Hiding the tanks and boilers in the roof rather than placing them in a tower on the roof was also related to this idea of being hidden at the back of the site.

Relationship with the Sea View

The site has sea views to the east and south and a mountain view to the west. It was important to consider the issue of views during the design of the building.

---

7 Yellow stone is a natural material for walls in the whole Cyprus. It was used for of the British period buildings. British authorities preferred this material to give a local appearance to the public buildings built by them. Currently it is usually used as a cover material.

As there are three empty plots of land between the RONAC's land and the sea, the sea view may become partially blocked in the future. Because of this she decided to be careful with the sea view.

> Since it is possible, I may lose the view in the future if the empty plots in front of mine are built on. I tried not to give extreme emphasis to the view directions. But still, I did not want to ignore them either. I directed myself towards the view. However, if they are closed to me in the future, I will still have a nice garden in front of me.

The RONAC also decided to have metal fences around the garden (except on the street side) so as not to block the view in any direction.

Reflection(s) from Theory

The fifteenth of the 17 SDGs suggests buildings should **include plants, insects, and animals**.[8] The RONAC preferred to have a natural garden and she had always lived with pets. Her new house was designed to accommodate many animals, such as cats, dogs, birds, and even black snakes ...

Security

The RONAC was not very aware of her desperate demand for security during the design process; however, many of her decisions were related to the need for security. Since she is a woman living on her own, these precautions were greater for her than for other people. The decisions related to security can be listed as follows:

- Buying a plot of land with firm soil type to avoid the worst of earthquakes.
- Fitting lockable shutters to all openings.
- Placing the dogs' kennels in the front section of the garden. The dogs are like alarm bells.
- Placing the house behind a dense group of trees as the RONAC thought this would reduce the bad effects of the two base stations 180 m away from the house.
- Deciding to keep a 10 m empty buffer area around the house. As the land does not have any other buildings around it, there is a danger of wildfires.
- Fitting decorative metal elements to the front of the windows with no shutters, for security purposes.
- Illuminating the garden and setting up some cameras, as the house is placed towards the back of the site.

---

8 United Nations, *An Architecture Guide to the UN 17 Sustainable Development Goals*.

I increased the number of solar lamps. All parts of the garden are well illuminated during the night now. They turn on automatically and when somebody goes near, the light suddenly becomes brighter. I feel secure because this can also be seen from inside. When the electricity is cut at night, I am never in dark because these lamps throw light inside too.

> *Reflection(s) from Theory*
>
> Tim Ingold compared a natural hill with architecture and wrote that any being can approach a hill; however, this is usually untrue for architecture. Architecture is distant to many beings including some humans but hills are close to everybody.[9] The Monarga House is not as **close to all beings** as the hill of Ingold. However, it is closer to many beings (cats, dogs, birds, etc.), except for human beings, in comparison to many other buildings and houses.

Colors

The RONAC was only partially conscious about the importance of colors to her during the design process. However, if one searches for the word "color" in her diary, one will find that color was very important at almost every step. Color was important during design (a white building, green shutters, etc.), during the application project (a preference for a lighter green for shutters to avoid the effect of dust), while selecting materials and systems (the color of ceramics, wardrobes, etc.), and during the construction of the building (deciding that the timber elements should be a lighter color). The RONAC preferred to be in a colorful environment to feel happier and refreshed. The selection of colorful ceramic tiles was especially a turning point for the later color preferences.

> So, I was very happy to choose this product and apply it everywhere in my house including terraces and footpaths towards my house. This tile also helped me to choose the colors of other surfaces inside and outside. I still feel that the floor covering is very important in a house. It determines the nature of the house. It is the surface you walk on. You can sit and look at it for a long time.

These ceramics had many colors in them and gave the RONAC the freedom to use colors, especially inside the house. This was also part of being open to all directions, nature, and all colors.

---

9 Tim Ingold, *Making – Anthropology, Archeology, Art and Architecture.* (Routledge, 2013).

Preliminary Design Ideas and Main Tectonic Decisions | **85**

Being Open to the Contributions of Builders, Foremen, and Workers

Construction of the Monarga House started in 2017. It was not a design decision open to the contributions of builders, foremen, workers; however, it happened spontaneously. Many builders, foremen, and workers suggested the inclusion of many details during construction and the RONAC accepted most of them. Many of these cannot be called changes because they relate to ambiguous issues. Contributions include the top detail of the stone walls, the places to apply leaf plaster, the organization of timber pieces in the timber screens and the design of the top part of the rain chain using timber. These changes were contributing to the architecture of the building.

The Inclusion of Tectonic Details

As the RONAC has taught tectonics courses for many years, she wanted to have some tectonic details in her house. One of these was the drainage channels under the passage near the entrance which existed in the preliminary project. However, most of the other tectonic details were designed during construction. Some of these were applied and others were not.

The design of the chimney cap became an important issue during the construction of the building; the suggestion in the preliminary project was not applicable due to the dimensions of the chimney. The chimney was much wider than the RONAC had expected. Then she decided to have a copper-colored chimney cap in the form of a witch's hat. She searched for some materials for this purpose.

> Today I called an architect friend and found that he did not use copper in his project. He used black iron. He rusted it by getting it wet. Then he applied a material called UV to protect the material. However, after applying this protective layer, the color changed slightly.

Eventually, the RONAC started worrying about the danger of a lightning strike attracted by a metal cap with a sharp point. The house is located 37 m above sea level, and she gave up on this idea due to fear. One of the builders designed a small four-sided pitched roof, using tiles for the chimney cap. Metal mesh was placed inside the chimney to stop birds getting in, and the top part of the chimney was painted dark gray to eliminate the sight of dirt or smoke on the white leaf plaster on the body of the chimney.

The front door was made of aluminum and had a side door (Figure 3.1b). However, the height of the door opening was more than the height of standard aluminum doors. It therefore became necessary to include a window above the door. The RONAC decided to have an ornamental metal element fitted to the front of this top window to deter burglars. However, when the aluminum firm delivered the front door, the RONAC found that the side door could not be opened. She could not accept this. Then the firm was in trouble because it was not possible for the aluminum painting firm to paint a single, small openable side door. That firm turns on their oven if there are many pieces to paint. So, the RONAC accepted a champagne colored aluminum side door with an openable window. Then it became necessary to have an ornamental metal element fitted in front of the top and bottom windows of this front door. The RONAC designed this metal element and another firm produced it towards the very end of the construction of the house. This detail can also be regarded as a tectonic detail.

Another tectonic detail relates to the two rain chains designed after completion of the construction because of a serious water problem caused by the roof. These rain chains were designed by some builders, a carpenter, and the RONAC. The photograph in Figure 3.2d shows the rain chain on the north façade.

The innovative dimension of these three details of the completed building can be questioned. The selection of the copper colored chimney cap was due to the innovative attitude of another architect. The front door details were not typical. It was unusual to use timber for the top pieces of the rain chain but it was necessary so as not to disturb the character of the timber fascia board at the end of eaves. However, none of these elements was designed during the preliminary design stage or the application project stage.

> *Reflection(s) from Theory*
> Three different sources highlight the critical role played by architectural details. Gevork Hartoonian wrote that details **relate analytically (disjoining elements) and poetically (considering the continuity of elements)** with each other.[10] Marco Frascari also related the technical and poetic dimensions by developing two different concepts: *logos of techne* (success in the technical dimension) and *techne in logos* (success in the **poetic dimension**). Both concepts cover the technical and poetic dimensions in their own ways.[11] Also, the philosopher Theodor W. Adorno asked for the simultaneous presence of **functionality and ornamentation in details** to avoid the **fetishist characteristic of functionless details**.[12]

---

10 Hartoonian, *Ontology of Construction*.
11 Marco Frascari, "The Tell-the-tale Detail." In *The Building of Architecture*. (1984). Accessed April 12, 2025. https://diffusive.wordpress.com/wp-content/uploads/2009/11/frascari-m-the-tell-the-tale-detail.pdf.
12 Theodor W. Adorno "Functionalism Today" *Oppositions* 17, 1979: 31–41.

> Yonca Hurol wrote about the **poetic roles of structural details, finishing details, and system details.**[13] Alvar Aalto's butterfly beams in the Saynatsalo Town Hall are examples of structural details. The window details of Tadao Ando's Church of Light are examples of finishing details. System details are architectural drawings which serve for the poetic purposes of achieving human scale, and designing places, corners,[14] and hiding places.[15]
>
> The chimney cap and front door details of the Monarga House as well as the rain-chain detail are attempts to combine the analytical and poetic dimensions and they are not functionless. The chimney cap and front door details are finishing details, while the rain-chain detail is a system detail which requires study of the plan, section, and elevation together.

Some Subconscious Dimensions in the Design of the Monarga House

Towards Minor Architecture

There were also some subconscious dimensions in this design process. These originate from a dislike of opulent buildings, a preference for not being identifiable, and a desire to have a building close to people and nature (in parallel with being open to all directions and everybody), which in time transformed the design into an example of minor architecture. But the RONAC became aware of this much later while furnishing the house.

> One important thing to record here is that I have not furnished the house with totally new things. I kept many old things that I love. I am very happy with that. Another important thing about furnishing is that I haven't placed the decorative elements to support certain styles in certain places. Instead, I have disturbed the formation of styles. I do not like perfection to be supported by the decorative elements. This is minor architecture. I also have all colors in the house. I do not like those white hygienic environments.

The design of the building also has some similar characteristics to minor architecture. It is not totally traditional, and it is not totally modern. It has some characteristics of both. Similar to having different design approaches and styles, having all colors in the building also had a representative dimension.

---

13 Yonca Hurol, *Tectonic Affects in Contemporary Architecture*. (Cambridge Scholars Publishing, 2022).

14 Christian Norberg-Schulz, *Genius Loci: Towards a Phenomenology of Architecture*. (Rizzoli, 1979). Edward Relth, *Place and Placelessness*. (Pion Ltd., 2008). Yi-Fu Tuan, *Space and Place: The Perspective of Experience*. 8th Edition. (University of Minnesota Press, 2001[1977]). Edward Casey, *The Faith of Place: A Philosophical History*. (University of California Press, 1998).

15 Bachelard, *The Poetics of Space*.

3 Preliminary Design of the House and Tectonic Affects due to Changes

I think I drew a picture and then I went into this picture. However, I still continue to draw and paint the picture. It is lovely. I feel like this ... One of my past students said that the colors of my house are lively. I agree with that. I think the ceramics gave me huge freedom about colors. Every color is there ... I just used my feelings, but it was not easy. It was a huge problem for me. I succeeded I think; However, there is a mixture of many styles. This is very much me. Combining everything and being open to everything. The house was designed to be open to four (all) directions. It is.

Reflection(s) from Theory

Minor architecture is political because it goes **against the authority formed through architectural mythologies**. It is about forming a *line of escape*[16] from the myths about the **separation between interior and exterior, architecture as a visible object, the architect as the subject, and nature as other**.[17] Each designer draws his/her line of escape.

The Monarga House was designed to have **continuity between indoor and outdoor spaces**. It can even be said that the indoor living area was designed to be transformed into a semi-open space. **The hidden character** of the house can be seen as a line of escape from the architecture as a visible object. **Mixing the styles** to avoid being identifiable and to be modest is a line of escape from the architect as a subject. **Not harming plants, having a naturally developing garden, and designing the house and garden for animals** are reflections that nature has not been excluded from the Monarga House.

Not Being Domestic

Breaking the domestic character of a house by removing the soft surfaces was another subconscious design tendency. Actually, the RONAC lived without such soft surfaces in her previous houses as well. However, she became aware of this situation in her new house after the completion of the building because of the comments of her friends.

---

16 The concept of line of escape (or line of flight) can be defined as the presence of a strong desire to find a way out of the cracks of the system and its authority. Gilles Deleuze and Felix Guattari, *A Thousand Plateaus, Capitalism and Schizophrenia*, Trans. Brian Massumi. 5th Edition. (Continuum, 2004[1980]). Gilles Deleuze and Felix Guattari, *Anti-Oedipus – Introduction to Schizoanalysis*, Trans. Eugene Holland. (Routledge, 2002[1972]).
17 Jill Stoner, *Toward a Minor Architecture*. (MIT Press, 2012).

Preliminary Design Ideas and Main Tectonic Decisions | **89**

> Then he said that I do not have any curtains. I said, no carpets either. I told
> him that I may have a few curtains at those places where there are problems
> between the doors and windows. He was worried about that idea. Then we
> started to talk about tectonics and the importance of soft surfaces: carpets,
> wall carpets, curtains etc., because humans can touch and lie on these sur-
> faces. I said that it is domestic. He said he does not like that. I said I also do
> not like that. So, in a way my house is a kind of place between an office and
> a house … that is me.

This non-domestic tendency originates from the presence of animals in and
around the house of the RONAC for many years. Currently, she has six cats at
home, four dogs, and more cats outside and around the garden. The cats destroy
carpets, furniture, and especially curtains. It is also difficult to clean carpets if there
are cats in the house. Therefore, it is a good idea to eliminate or minimize the use
of carpets and curtains and to use non-destructible furniture. Similar things can be
said for the garden. It is not possible to have an elegant garden with comfortable
outdoor furniture if you have many animals in the garden. Living with animals
requires the acceptance of non-comfortable furniture and objects. In any case, the
RONAC does not like to get too relaxed. She thinks it is better to be open to the presence of nature and to live with nature. This is also a part of being open to the four
cardinal directions.

Shutters are necessary for animals. Shutters are very helpful when opening
windows without losing control of animals. The RONAC had thick walls built
for her house and she decided to leave the maximum space between the window
frames and the shutters, so that the cats can sit, run, and use these narrow spaces.

> The windows and shutters work very nicely for cats. I open windows in the
> daytime and close them at night. Poko likes to sleep between the windows
> and the shutters. I am happy about them.

The RONAC wanted an open-plan house during the first phases of the preliminary design. She thought about having one main space divided by wardrobes,
cupboards, bookshelves, or screens. However, she gave up on this idea because
of the animals. Having animals in and around the house requires closable rooms.
If an animal gets ill, or an animal is rescued, it is necessary to put that animal in a
separate room. If you have to give medicine to an animal, you have to separate that
animal from the others. If you prepare some special food for one of the animals,
the kitchen should be closable. It is easier for cats to have their own safe corners
if the house has separate rooms.

> *Reflection(s) from Theory*
>
> It can be stated that this goes against the tectonics theory of Gottfried Semper which praises the presence of **soft surfaces** as being close to humans.[18] The Monarga House has hard surfaces. It has very few rugs and there are no curtains.
>
> The fifteenth SDG suggests buildings should **include plants, insects, and animals.**[19] The Monarga House is open to many plants and animals.

Memories – The Invisible

If an architect designs her own house, there can also be an unconscious dimension behind this. It is not possible to identify all unconscious dimensions of this process without a psychologist; however, while analyzing the diary of the RONAC, and considering certain notes about her childhood memories, it became essential to relate them to the design ideas of the Monarga House. The paragraphs below were written four months before the RONAC moved into her new house.

> I was looking for my mother's childhood house in Famagusta. Then my friend's mother said that she knew of a house which had been used by judges in the old colonial days of Cyprus and she showed me that house. I liked it and I took photos. Then I sent the photos to my son, and he replied saying that "this is very similar to your house" (my new house). I looked at the photos again and was surprised to see that my original project was similar to this building. I had white garden walls topped with red tiles. I had dark green timber elements. These were things I had to change in time. I changed the dark green because my friend said it would get dusty. I changed the color of some timber members because they were not sufficiently thick. I even started to imagine the future upper floor of the house in a similar way. The large garden, the greenery, and red roofs ... it is the same image. I wrote about this to my friend who works on meaning in architecture, and she said that she was shocked.

> I was shocked because I saw this house for the first time ... The large rooms; the high ceilings; I know this much ... I am not sure if I have seen any photos of it or not ... Maybe I did. Maybe my mother took a photo of this house and put it into one of her suitcases of photos. In any case I designed my house totally subconsciously when its relation to this "memory" is considered. I remember that I saw a little house in Boğaztepe at the edge of the abyss and in greenery and I was inspired by that house.

---

18 Gottfried Semper, *The Four Elements of Architecture and Other Writings*, Trans. Harry F. Mallgrave and Wolfgang Herrmann. 2nd Edition. (Cambridge University Press, 2010[1851]).
19 United Nations, *An Architecture Guide to the UN 17 Sustainable Development Goals.*

Figure 3.4 (a) The small house in Monarga which inspired the project. (b) The RONAC's mother's childhood house in Famagusta.

Was I attracted to this house because it is similar to my mother's childhood house and it brings back the stories she told me? Probably.

> Even before seeing this house in Monarga, I was trying to buy houses with a pitched roof and timber elements. Step by step I ended up here ... I told my friend long ago that I'd become attached to Cyprus before coming here because of the stories and photographs of my mother. If there is no photo of this house, then this shows the strength of the imagination to construct a totally correct mental image of a house on the basis of words alone.

The small Greek Cypriot house, which was a source of inspiration for the design of the Monarga House, and the RONAC's mother's childhood house can be seen in Figure 3.4.

Tectonic Characteristics of the Completed Building

Since the RONAC is an academic who studies tectonics in architecture, it can be expected that these theories also inspired her during the design phase of the Monarga House. The tectonics of the Monarga House can be investigated according to:

- Holistic tectonics which considers all the physical factors that affect architecture
- Theories of tectonics
- Theory of tectonic affects.

Since a holistic investigation of the tectonics of a building covers all physical issues about that building, an analysis of photos of the completed building and the design ideas and decisions of the Monarga House reveals that the building materials and details in this building play an important role in its tectonics, in addition to the tectonic use of the topography and the tectonic affects of climate on the building.[20] Tectonic use of materials was less important during the project stage. However, it became more important because of the addition of stone and timber elements and components to the building during the construction stage. Details also became more important during and after the construction phase. Fireplace and chimney details existed in the preliminary design project; however, the rain-chain detail became possible only after the construction process. The important role of the topography and the site and the climatic concerns of the design project were followed up during construction and the completed building has most of these tectonic characteristics. Still, it can be said that considerations of the topography further increased during construction. Some people might also think that this building also considers the tectonics of Cypriot culture because of the vernacular look of some façades and the use of local leaf plaster. Table 3.1 demonstrates the results of the holistic investigation of tectonics in the Monarga House.

An evaluation of the Monarga House, according to influential tectonic theories, first requires a consideration of Karl Botticher's concepts of *Kernform* and *Kunstform* (in which structure becomes ornament and ornament becomes structure).[21] Since the Monarga House is composed of a reinforced concrete frame structure and brick walls covered with leaf plaster and paint, it does not

Table 3.1 Holistic investigation of tectonics in the Monarga House – an outline of the reflections from theory.

Categories covered by holistic tectonics	**Materials** (developed further during construction)
	Structural system
	Mechanical system
	Electrical system
	Information technology
	Details (developed further during construction)
	Light
	Topography (developed further during construction)
	Climate
	Culture

The bold categories in the table show the Monarga House's characteristics.

---

20 Yonca Hurol, *The Tectonics of Structural Systems – An Architectural Approach.* (Routledge, 2016).

21 Karl Gottlieb Wilhelm Botticher, "The Principles of the Hellenic and Germanic Ways of Building with regard to their Application to our Present Way of Building." In *In What Style Should We Build?* Eds: J. Bloomfield, K. Forster and T. Reese. (The Getty Center Publication Programs, 1992[1828]): 147–167.

align with Botticher's tectonic values, as the structure is not visible. However, this implies that all modern buildings with covered frame systems and walls do not conform to this approach.

When applying Gottfried Semper's concepts of *tectonics* (lightness) and *stereotomics* (heaviness) to the Monarga House,[22] it can be argued that the house embodies a stereotomic approach, given its reinforced concrete structure and numerous brick walls. The Monarga House does not appear light, and therefore, it does not align with Semper's tectonic approach.

Applying Eduard Sekler's concepts of *tectonic* and *atectonic* to the Monarga House[23] also yields a negative assessment regarding the building's structural system and walls. This is because the reinforced concrete structure and walls are concealed behind leaf plaster and paint in an atectonic manner. However, other elements and components, such as the roof and pergolas, exhibit a tectonic character, as they largely reveal the reality of their construction. Consequently, the Monarga House can be considered partially tectonic and partially atectonic according to this approach.

Kenneth Frampton's concept of *tectonic form* does not solely focus on the structural system.[24] The Monarga House exhibits a partial tectonic form because it partially reveals its construction logic. However, since the logic of construction is not fully visible and understandable, the building's structure lacks *autonomy* according to this approach, which means the Monarga House is not *ontologically grounded* from a tectonic perspective. It is also possible to describe it as *scenographic*, given that the structure and walls are covered with leaf plaster, making the building resemble some vernacular examples with masonry walls and leaf plaster.

Gevork Hartoonian's concept of an *ontological approach to time* applies more appropriately to the Monarga House.[25] There are historical building elements and components in this house, such as leaf plaster, the shutters, the pitched roof, but these elements have not been reinterpreted specifically for this house. Notably, a modern metal fireplace prominently features in a central position, and the size of the openings was determined according to the functions of the different

---

22 Semper, *The Four Elements of Architecture and Other Writings*; Gottfried Semper, *Style in the Technical and Tectonic Arts; or, Practical Aesthetics,* Trans. Harry F. Mallgrave. (Getty Research Institute, 2004[1860]).

23 Eduard Sekler, "Structure, Construction, Tectonics." In *Structure in Art and Science*, Ed: G. Kepes. (WordPress, 1965): 89–95. accessed August 18, 2018. https://610f13.files.wordpress .com/2013/10/sekler_structure-construction-tectonics.pdf.

24 Brian Kenneth Frampton, "Rappel a l'Ordre: The Case for the Tectonic." In *Labour, Work and Architecture*, Ed: K. Frampton. (Phaidon Press, 2002): 91–103. Brian Kenneth Frampton, *Studies in Tectonic Culture – The Poetics of Construction in Nineteenth and Twentieth Century Architecture.* (MIT Press, 1995).

25 Gevork Hartoonian, *Ontology of Construction – On Nihilism of Technology and Theories of Modern Architecture.* (Cambridge University Press, 1994).

spaces. The living area has large openings similar to a modern building, while the other spaces have smaller and fewer openings, akin to traditional buildings. This approach shows similarities to Adolf Loss' ontological approach to time, as described by Hartoonian. Loss used small openings for domestic functions to ensure privacy in the Losshouse, while much larger openings were used to reveal the presence of a frame system for public functions. The use of a modern fireplace in a central location in the Monarga House, similar to how Frank Lloyd Wright incorporated it in his Prairie architecture, could also be seen as a sign of an ontological approach to time, as this fireplace is modern and open on two sides.

Gottfried Semper's concept of *hearth* is applicable to the fireplace and the surrounding living space of the Monarga House.[26] The fireplace plays a central role in the house. Many functions are organized around it and it marks the main living (not gathering) area. However, soft, touchable materials and surfaces emphasized by Semper are absent from the Monarga House. In fact, the opposite seems to have been the aim. Most of its surfaces are rough, including the leaf plaster, the lack of curtains and carpets, and the presence of stone walls. The RONAC describes this as a deliberate choice to avoid creating a domestic atmosphere. This idea began during the design phase and continued to evolve throughout the construction and use phases of the building.

Gevork Hartoonian's concept of *montage* (joint-disjoint)[27] also applies to the Monarga House, as the house simultaneously combines two different approaches – traditional and modern – in its configuration. One example is the use of small openings in the rooms (for climatic comfort), which create a masonry-like appearance, contrasted with the large openings in the living area (being open to four directions) that clearly express its modern nature with its frame system. Therefore, the Monarga House embodies the concept of montage. Table 3.2 illustrates the characteristics of the house according to these theories of tectonics.

If the tectonic affects[28] which are caused by the Monarga House are studied, it can be stated that *poetic tectonic affects* are dominant because of the haptic character of the leaf plaster and the presence of tectonic details. The haptic character started during the design phase with the use of leaf plaster and timber pergolas and increased during construction because of the addition of the stone and timber parts to the building. There were some tectonic details in the design project, such as the fireplace and chimney; however, the type and number of tectonic details increased during the construction phase. Rain chains can be considered as one of these details. There are also some corners and places in and around the house which include the study area, dining area, and the small

---

26 Semper, *The Four Elements of Architecture and Other Writings*.

27 Hartoonian, *Ontology of Construction*.

28 Hurol, *Tectonic Affects in Contemporary Architecture*.

Table 3.2 Investigation of tectonic characteristics of the Monarga House according to influential theories of tectonics – an outline of the reflections from theory.

Some evaluation/understanding categories in the influential theories of tectonics	Botticher's Kernform and Kunstform
	Semper's tectonics versus **stereotomics**
	Semper's earthwork, **hearth**, framework, and enclosing membrane
	Semper's **soft surfaces** (the opposite was achieved because of the design) and knots
	Sekler's **tectonic** (except the structure) versus **atectonic** (structural system)
	Sekler's balanced image and its relation to structure
	Frampton's **tectonic form** (except for the structure); being ontological versus **scenographic**
	Hartoonian's **ontological approach to time**
	Hartoonian's **montage (joint-disjoint)**
	Hartoonian's balance between technological and artistic

The bold categories in the table show the Monarga House's characteristics.

north and west terraces. Continuity with the natural context is a main poetic feature of the building. However, the early design intentions of being hidden at the back of the garden and creating a small building were not fully realized because of the increase in the floor height due to climatic concerns. These can also be interpreted as intentions towards the poetic tectonic affect of the "house and universe"[29] which has been partially realized. The building design has an affirmative approach to the tectonics of change which means that there is no technological improvement/innovation in the design of this building. The building looks familiar rather than new. It is possible to say that the building reflects an ontological approach to time as explained in the previous paragraph. The design intentions verge towards a building which is close to people (as a "thing" and not as an object) because it is not dominant. Simplicity, and not being ostentatious make this building a thing rather than an object. Thingness was also partially achieved because the increase in the floor height effected the image of the building in the opposite direction. The tectonic affects caused by the Monarga House are presented in Table 3.3.

29 Bachelard, *The Poetics of Space.*

96 | *3 Preliminary Design of the House and Tectonic Affects due to Changes*

Table 3.3 Tectonic affects caused by the Monarga House – an outline of the reflections from theory.

Poetic tectonic affects	Tectonic affects of change	Tectonic affects of time	Tectonic affects due to domination
• Lightness/heaviness • **Hapticity** (developed further during construction) • **Tectonic details** (developed further during construction) • Human scale in detail • **Corners/places in detail** • Hiding places in detail • **Continuity with the natural context** (partially realized because of the design) • Continuity with the historical context • Continuity with the urban context • **House and universe** (partially realized because of the design)	• Technical improvement • Innovation(s) • **Affirmative approach** • New/**familiar** • Practicality	• Timeliness • **Ontological approach to time** • Rejection of history • Conservatism • Futurism	• Expression of power • Territorialization • Aestheticization of politics • Culture industry • Sensational image making • Domination of form • **Thing** (partially realized because of the design)/object • Precision/imprecision

The bold categories in the table show the tectonic affects caused by the Monarga House.

These explanations and tables give an impression that the phases of the application project, tendering process and construction of the Monarga House were the sources of some tectonic achievements. However, the following chapters of this book demonstrate that there were also important losses and gains during construction of the building and it was like warfare. The first phase after the preliminary project phase is the application project phase.

Tectonic Affects and Innovative Attitudes due to Changes During the Application Project

The application project was drawn up by an experienced and well-known architectural firm owned by a friend of the RONAC. Since the RONAC designed the

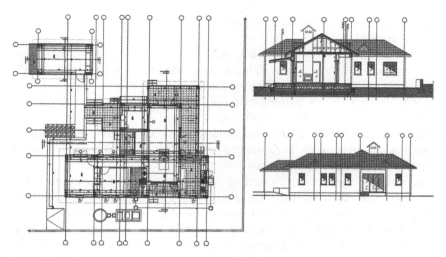

Figure 3.5 Some drawings from the application project.

preliminary architectural project, she has also been accepted as the architectural project designer (*muellif* in Turkish) for the application project. The RONAC met with the manager of this firm several times to answer questions, to approve the changes made, and to approve the final application project.

One drawing sheet from the application project is shown in Figure 3.5. This drawing can be useful in identifying some of the changes by a comparison with the preliminary project shown in Figure 3.1.

Changes During the Application Project

- The distance between the house and the site borders was increased to 4 m from 3 m to be on the safe side with respect to regulations. This caused the loss of two trees at the front of the house. Also, the house became too close to some other trees. The biggest of these trees collapsed and was later removed. However, this move made it easier to have new trees between the fence and the house.
- One more window was added to the small bedroom. Considering that the room faces west, there was only one small window in that room in the preliminary design project. However, the architect of the application project said that it would be too dark and because of this one more window was added to that room. This affected the building positively by breaking the very simple and monotonous order of the back façade.
- The bedrooms were enlarged and accordingly the size of the bathroom was reduced. While doing this the wardrobes in these bedrooms were also reduced to gain more space for these rooms. This resulted in more spacious bedrooms

which can also accommodate double beds. The study and dining areas were also enlarged.

- The organization of the bathroom and kitchen was also changed because the architectural firm was working in coordination with the firms providing the kitchen and bathroom furniture and equipment. The washing machine was removed from the bathroom to the kitchen to decrease the size of the bathroom and enlarge the two bedrooms.
- One of the two beams dividing the living area into three was removed during the preparation of the civil engineering project. This provided better continuity between the two parts of the living area.
- The fireplace was moved forward to open up the space for a dining table. However, this disturbed the living area because the fireplace was blocking two of the large openings; this move had to be reversed during construction.
- The RONAC wanted to have accordion windows for the large openings in the living area. However, the architect who dealt with the application project warned her about their poor quality of construction which affects comfort in houses. The sliding window type was then chosen to avoid this problem. However, this made an impact on the tectonic affects because the continuity of the living area with the garden was decreased. Also, the sliding shutters had to be changed for the same reason.
- The floor level of the indoor spaces was the same as the floor level of the terraces in the preliminary design project to increase the continuity between the indoor and outdoor spaces. However, the application project was changed and the level of the indoor floor was made higher than the outdoor floor level so that water could not enter. This decreased the continuity between the indoor and outdoor spaces as well. However, in any case, there would also be the thick rails of the sliding windows which also separate the two spaces.
- The application project architect wanted to connect some of the terraces to each other. However, the RONAC did not accept this.
- The roof gutters which existed in the preliminary project, were removed and the roof was designed to have free falling water. This caused some positive tectonic affects because of the elimination of rain gutters from the façades. However, one of the reasons for adding pavements around the building later was also related to the removal of the gutters and those pavements do not look nice.
- Traditional roof tiles were replaced with modern red roof tiles to decrease the cost of the building.
- The ceramics were chosen during the application project. Colorful tiles which represent heterogenous people in Istanbul inspired the RONAC and she made a selection in that direction. This resulted in having many colors on the floor which gives her the freedom to use many colors in the interior spaces. Tiles with a traditional appearance were chosen so as not to be too ambitious.

- As the civil engineering project was finished before the completion of the application project, the type of foundations was decided and this was reflected in the drawings of the application project. Since this was not visible, it did not cause any tectonic affects later.
- When the mechanical engineering project was done, the placements of the water tank, septic tanks, and so on, were determined. However, the water tank was not the right size because a 4-ton tank was requested by the RONAC and a 2-ton tank was included in the application project. This had to be changed during the construction process of the building. If the size of the tank had not been changed, this would have caused a water shortage in a house with many animals. The location of the tank was also changed later because it was in front of the second bedroom in the application project. This was eliminated during the construction process by putting a 5-ton tank in front of the bathroom.
- The electrical engineer placed the ACs without asking the RONAC and she missed seeing their locations when she approved the application project. Finally, one of the AC boxes is on the best terrace (south) of the house.
- The central heating system heating the floors was eliminated to decrease the cost.
- Instead of applying heat insulation to the walls, it was decided to use eco-bricks during the application process. This resulted in heat insulation only on the reinforced concrete frame elements but not on the walls.
- There was a contradiction between the preferences of having heat insulation on the reinforced concrete structural elements and the use of leaf plaster. The leaf plaster is heavier than normal plaster and the heat insulation layer cannot carry its weight. However, it eventually became possible to have the leaf plaster, because the heat insulation on the columns and beams became impossible during the construction stage.

Reflection(s) from Theory

Some of the changes made during the application project phase were done to **increase the commodity value**[30] of the Monarga House. The preliminary design project was designed for the RONAC alone. However, the architect of the application project questioned such issues and suggested having double beds in both bedrooms and a large dining table in the living area. He thought these changes would increase the value of the house as well as helping different owners of the house in the future. The RONAC accepted these suggestions.

Some other changes originated from the architect of the application project regarding the **low quality of construction** in North Cyprus.

(Continued)

---

30 Judith O'Callaghan "Architecture as Commodity, Architects as Cultural Intermediaries – A Case Study" *Architecture and Culture* 5, no. 2, 2017: 221–240.

> **(Continued)**
>
> These included the types of sliding doors and shutters, and no difference in level between the indoor and outdoor environments. These suggestions reflected being on the safe side; however, they also eliminated the **possibility of innovation** during construction. Being on the safe side might decrease innovation and eliminate the possibilities for a **constantly evolving architecture** towards a better functionality and aesthetics.[31] This also has negative consequences for the evolution of the building market. However, the RONAC accepted being on the safe side because she was worried about the cost of her house and future complications which could arise.
>
> The changes in the engineering projects indicate a **lack of collaboration between the professionals** and the client during the application project phase.[32] This caused various problems later.
>
> Some other changes were done to decrease the cost of the house.[33] **Affordability** was a major concern for the RONAC.

The application project made an important contribution to the architecture of the Monarga House. The list above makes this contribution obvious. However, the tectonic affects as a result of this process were not as positive. The RONAC believes that these tectonic problems arising from the application project were due to her quick inspection.

> The project production was so slow that I was checking it very quickly and because of this I did not see many of the problems originating from the application project. I missed many things in that way and saw these problems during construction when it was too late.

Tectonic Affects due to Changes in the Application Project

Some of the changes made during the application project caused tectonic affects and some of them did not. Tectonic affects due to changes in the application project are listed in Table 3.4.

---

31 Christos Chantzaras, Architecture and Design of Innovation Processes. PhD Thesis from Technische Universitat München, TUM School of Engineering and Design, 2022, accessed May 28, 2025. https://www.researchgate.net/publication/364353696_Architecture_and_Design_of_Innovation_Processes_-_Applying_architectural_thinking_and_tools_to_the_understanding_and_design_of_innovation_processes_in_innovation_management.
32 Yonca Hurol "Ethical Considerations for Designing Buildings with Reinforced Concrete Frame Systems in Earthquake Zones" *Science and Engineering Ethics* 20, no. 2, 2013: 597–612.
33 E. Mueller and R. Tighe (eds.), *The Affordable Housing Reader*. 2nd Edition. (Routledge, 2022).

Table 3.4 Tectonic affects due to changes during the application project.

The change in 1a	The symbol	The tectonic affect(s)	Responsible person/contributor	Related changes	Innovative attitude
A window was added to the west façade.		This broke the monotonous character of the façade and provided better continuity.	Architectural firm, the RONAC.	0	None
The fireplace was moved forward to open up more space for the dining area.		Since the fireplace was blocking the view and cross ventilation came from two directions, this ran counter to one of the main design decisions about continuity.	Architectural firm. The RONAC accepted this without being aware of it.	Many	None
Accordion windows and sliding shutters were eliminated because of the bad quality of construction in North Cyprus.		Since this was decreasing the feeling of openness of the living area, it ran counter to one of the main design decisions about continuity.	Architectural firm, the RONAC.	Many	Yes but not realized.
One of the two beams dividing the living area was removed.		This removal provided better continuity in the living area.	Architectural firm, the RONAC.	0	None
Use of different floor levels for the terraces and the interior spaces to eliminate water coming in.		Since this was also decreasing the feeling of openness of the living area, it ran counter to one of the main design decisions about continuity.	Architectural firm, the RONAC.	0	Yes but not realized.

(Continued)

Table 3.4 (Continued)

The change in 1a	The symbol	The tectonic affect(s)	Responsible person/contributor	Related changes	Innovative attitude
Connecting the terraces to each other was eliminated.		This would change the presence of different terraces in different directions and decrease the natural character of the project.	Problem: Architectural firm Solution: the RONAC.	0	None
Traditional roof tiles were replaced with modern ones.		This was necessary to decrease the cost, but negatively affected tectonics about continuity with historical buildings.	Solution: Architectural firm, the RONAC.	0	None
Roof gutters were not fitted, and the roof was designed for water to fall freely.		This eliminated metal pipes on the façade and created a better tectonic detail. However, it became necessary to add pavements and two rain chains to the building.	Architectural firm, the RONAC.	Many	None
Change of the tank capacity from 4-tons to 2-tons.		This was functionally not acceptable and changed again during the construction phase. This caused RONAC to have a large and circular 5-ton tank later.	Problem: Architectural firm The RONAC accepted the 2-ton tank without being aware of it Solution: Rectified later by the RONAC	Many	None

The placements of ACs were not determined in an architectural way.		One of the ACs was placed on the south terrace with a sea view and obstructs the opening of one of the shutters. This is a bad tectonic detail.	Problem: Architectural firm The RONAC accepted this without being aware of it.	0	None
The contradiction between the heat insulation on the structural elements and the leaf plaster was left ambiguous.		This could have canceled the application of leaf plaster and would have been a negative tectonic affect.	Problem: Architectural firm Accepted by: the RONAC.	Many	Yes, but not realized.

Eliminated changes/problems are shaded in light gray in the table. Changes which cause another change or originate from another change are shaded in darker gray. Changes that did not cause any other changes are unshaded. Symbols reflect not only the type of change, but also the associated feelings. Lighter colors in symbols reflect positive feelings.

104 | 3 Preliminary Design of the House and Tectonic Affects due to Changes

Four of these are multiple changes and they made an impact on the tectonic affects caused by the building and because of them there had to be some series of changes during the various stages of construction. The RONAC's diary contains a note about her thoughts on the problem of the heat insulation on the structural system and the leaf plaster, which is one of these multiple changes.

> It might be necessary to increase the strength of heat insulation surfaces by applying some mesh to them, but it is possible now. Happy ending. Back to white walls again.

This approach left this as an ambiguous issue in the application project. Since the presence of heat insulation affects leaf plaster, and leaf plaster is better painted white, there is a chain reaction between these decisions. The RONAC believed that the problem could be solved by increasing the strength of the plaster by applying mesh over it. However, this did not work.

There were three innovative attitudes related to tectonic affects during the application project phase. One of them involved the heat insulation layer and leaf plaster, while another focused on the use of accordion windows, sliding shutters, and maintaining the same floor level between indoor and outdoor spaces.

Although the RONAC asked the chief architect of the architectural firm for a 4-ton tank, the mechanical engineering project only budgeted for a 2-ton tank and the RONAC was not aware of this change when she controlled the project. Finally, she had to insist on the reversal of this change, and she was given a 5-ton tank. The next year, after moving to the house, she wrote the following statement in her diary:

> At the end of summer North Cyprus had a serious water shortage. I am now so happy that I insisted on having a 5-ton tank. Otherwise, this would have been big trouble for me with four dogs and many cats.

Innovative Attitudes in the Application Project Phase

There are four issues relating to the innovative attitude of the application project phase of the Monarga House which can be discussed. These encourage innovative attitudes because they are challenging technical problems and solving them requires technical knowledge and experience.

- The continuity between the indoor and outdoor spaces required the use of accordion windows and sliding shutters. However, these were changed during the application project phase with the approval of the RONAC because of the bad quality of construction which causes problems with comfort. This change was against the innovative attitude and the outcome caused a negative tectonic affect.

- Second, having the same indoor and outdoor floor levels was also a requirement for providing continuity between the indoor and outdoor spaces. This was also changed with the approval of the RONAC, and this also ended the possibility of having an innovative attitude. The outcome of this change also caused a negative tectonic affect.
- There was a contradiction between the application of heat insulation layers on the reinforced concrete structural elements and having leaf plaster on all surfaces, because the heat insulation layers cannot carry the heavy leaf plaster. The ambiguity of this issue could have supported an innovative attitude later. However, this did not happen because of doubts about the strength of the heat insulation layer. Solving the heat insulation problem was not pursued. Still, the outcome caused a positive tectonic affect because of the aesthetic character of the leaf plaster.

> *Reflection(s) from Theory*
> Can ambiguity in architectural projects be suggested as a reason for innovation? According to Mahairi McVicar, the function of architectural projects to translate ambiguous qualities of architecture into a quantifiable and precise medium to enable construction of those buildings needs careful consideration. Although the intention is to provide perfect documents, there will always be some **ambiguity** in application projects.[34]

Tectonic Affects and Innovative Attitudes due to Changes During the Tendering Process

The RONAC contacted four construction firms for tendering purposes. Three of them saw the project and made contributions to the building by suggesting some reasonable changes. These changes were as follows:

- The need for a drainage well because of the slope and position of the site with respect to the pavement. This change was realized. It does not cause any tectonic affects because the well is in one of the far corners of the garden and it does not have any construction above it.

---

34 Mahairi McVicar, *Precision in Architecture – Certainty, Ambiguity and Deviation.* (Routledge, 2019).

- The need to have a pool under the water tank and boiler in the attic for water isolation purposes in case of any flooding. This need was contradictory to the requirements of the cold roof because cold roofs have heat insulation layers on the ceiling which is under the attic. This triggered many changes in the later stages of the building and the target of the cold roof has been partially achieved. It does not cause any tectonic affects because it is not visible.
- A change in the timber cover under the eaves of the roof and pergolas was needed because the application project contained an aesthetically unacceptable material. This change was realized, and it caused positive tectonic affects.
- A change to the connection detail of the timber pergola posts to the floor was necessary. The detail in the application project was not correct. This change was realized, and it caused positive tectonic affects.
- Stepped reinforced concrete tie-beams under the garden fences were suggested. This change was also realized. It caused positive tectonic affects.

Tectonic Affects due to the Changes in the Tendering Process

Three of the changes during the tendering process caused tectonic affects (Table 3.5).

Table 3.5 Tectonic affects due to the changes during the tendering process.

The change in 1b	The symbol	The tectonic affect(s)	Responsible person/ contributor	Related changes	Innovative attitude
Change of the timber cover under the eaves of the roof and pergolas.		This change initiated the use of timber in the building and its haptic affect.	Contractor	Many	Yes, realized.
Change of the connection detail of the timber pergola posts to the floor.		A special visible joint detail was used to connect posts to the floor.	Contractor	0	Yes, realized.
Having stepped reinforced concrete tie-beams under the garden fences.		This initiated the use of factory produced fences which look more transparent and lighter.	Contractor	Two	No, because it is typical.

Eliminated changes and problems are shaded in light gray in the table. Changes which caused another change or originate from another change are shaded in darker gray. Changes that did not cause any other changes are unshaded. Symbols reflect not only the type of change, but also the associated effects. Lighter colors in symbols reflect positive feelings.

Innovative Attitudes During the Tendering Process

Most of the five changes suggested by the contractors during the tendering process were innovative. The ones with tectonic affects were suggested by the contractor who carried out most of the construction of the Monarga House after signing a turnkey contract with the RONAC. Two of the tectonic changes during this phase also incorporated innovative attitudes.

> *Reflection(s) from Theory*
> Technical innovations in low-cost housing design have been on the agenda of architecture for a long time. There were some publications about affordable[35] and innovative buildings in the twentieth century. The Home House Project presents a competition's results in respect of some technically innovative designs towards **affordability**.[36] There is some contemporary research on social innovations in affordable housing including collaborative and participatory approaches.[37] Some other examples of contemporary scientific literature on innovations are related to artificial intelligence, collaboration, parametric design, sustainability, 3D printing, construction robots, and the use of new materials.[38] Innovative attitudes during the tendering process of the Monarga House related to **small and unsystematic changes** based on technical knowledge and experience of the contractor in respect of common building materials, details, and building techniques. However, this type of innovative attitude can also relate to the use of new materials and systems, and they can serve sustainability, affordability, and so on, in an unsystematic way.

Conclusion

The Monarga House is a small, simple house designed to combine modest architectural and tectonic qualities. It utilizes materials to achieve hapticity, maintains continuity with nature, leverages topography, and considers climate and culture. Additionally, it features tectonic details such as leaf plaster. The preliminary design did not exhibit any innovative tendencies.

---

35 National Housing Bank, *Low Cost Shelter Design Options*. (National Housing Bank, 1995).
36 D.J. Brown (ed.), *The Home House Project: The Future of Affordable Housing*. (MIT Press, 2005).
37 Grard Van Bortel et al., *Affordable Housing Governance and Finance – Innovations, Partnerships and Comparative Perspectives*. (Routledge, 2019).
38 S. H. Ghaffar et al., (eds.), *Innovation in Construction – A Practical Guide to Transforming the Construction Industry*. (Springer, 2022).

108 | *3 Preliminary Design of the House and Tectonic Affects due to Changes*

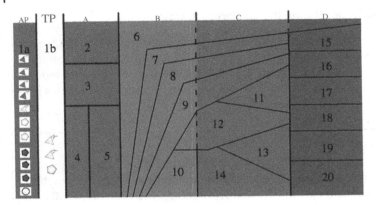

Figure 3.6 The map of tectonic process for the application project (AP) and tendering process (TP).

Many changes were made during the application project (AP) and tendering process (TP). Figure 3.6 shows the map of the tectonic process for these stages.

As seen in Figure 3.6, many changes occurred during the application project phase, resulting in both positive or negative tectonic affects. The main intention was to increase the commodity value. While some changes had positive tectonic affects, others had negative tectonic affects, even though they were functionally or structurally useful. The most important feature of the application project phase is the presence of changes that trigger other changes. The tendering process resulted in only positive tectonic changes, with two of them triggering further changes.

Table 3.6 shows the number of changes that occurred during the initial phases of the Monarga House and the innovative attitudes these phases fostered. This table allows for a comparison between the total number of changes (including changes causing tectonic affects and hidden changes), tectonic changes, and innovative attitudes. Although some innovative attitudes encompass changes that are not visible, this comparison provides insight into the general trend during these phases. The application project phase of the Monarga House combined many changes with tectonic affects and included some unrealized innovative possibilities. These issues remained ambiguous within the project and were not addressed later. In contrast, the tendering process involved a few changes causing considerable tectonic affects (some of which triggered many additional changes), but all innovative attempts within this phase were realized.

The first part of this chapter, focusing on the preliminary project of the Monarga House, includes various theoretical reflections from tectonics theory, architectural theory, philosophy, and sustainability theories. The second part of the chapter,

Table 3.6 The changes during the application project and tendering process of the Monarga House.

The phases AP, TP	Eliminated changes	Singular changes	Multiple changes	Total number of tectonic changes	All changes	Innovative attitudes
Application project (AP)	1 positive	2 positive, 3 negative	1 positive, 4 negative	4 positive + 7 negative: 11	18	3 unrealized and caused tectonic affects.
Tendering process (TP)	None	1 positive	2 positive	3 positive	5	5 realized, 2 with tectonic affects.
Total	1 positive	3 positive, 3 negative	3 positive, 4 negative	7 positive + 7 negative: 14	23	8 possibility but only 5 realized. 5 with tectonic affects but 2 realized.

which addresses the changes during the application project and tendering process phases, also contains theoretical reflections on architectural and construction theory.

Reflection(s) from the theory about the preliminary design phase contained some **tectonic** keywords such as métier, relating analytical and poetic, logos of techne and techne of logos, the poetic role of structural details, finishing details, or system details, soft/hard surfaces, montage, tectonics through materials, details, topography, climate, culture, stereotomics, hearth, atectonic, scenographic, affirmative approach, familiarity, thing, ontological approach to time, hapticity, corners, places, hiding places and continuity with the natural context.

Some **architectural** keywords such as Prairie architecture, accessibility, affordability, minor architecture, against the authority formed by architectural myths, separation between interior and exterior, architecture as a visible object, architect as the subject, nature as other, and mixing the styles.

Some **philosophical** keywords such as house and universe, close to all beings, functionality and ornamentation in details, and fetishist role of functionless details.

Some keywords from **sustainability** such as passive heating, cooling and ventilating, elimination of loss of vegetation and biodiversity, including plants, insects, and animals, and accessibility.

> *Reflection(s) from the theory about the application project phase* contained some **architectural** keywords such as increasing the commodity value, constantly evolving architecture, affordability, and ambiguity.
>
> Some **construction** keywords such as low quality construction, possibility of innovative attitudes, and lack of collaboration between professions.

> *Reflection(s) from the theory about the tendering process* contained one **construction** keyword which is small and unsystematic changes.

Part II

Changes in the Construction Process of the Building and Tectonic Affects/Innovative Attitudes

Part II demonstrates the process of construction of the Monarga House with the help of the diary and observation notes of the RONAC and the photo collection, and separately lists the changes which have happened during different phases of construction. It explains the tectonic affects caused by these changes and classifies them. It also identifies the innovative attitudes within these changes. This leads to an interpretation of the possibilities of the building which have been avoided as well as the tectonic achievements and losses caused by these changes. Chapter 4 contains the tectonic affects due to changes during the construction of the foundation system, frame system, walls, heat insulation, openings, and roof. Chapter 5 focuses on the phases of plastering, water isolation, ceramics, and mechanical and electrical systems. Chapter 6 covers the stages of closure and finishing. Chapter 7 is about the construction work in the garden. Each chapter ends with a process map for tectonic affects and a table presenting the innovative attitudes, which also allows for an introduction to their impact on the building.

Tectonics as a Process in Architecture, First Edition. Yonca Hurol.
© 2025 John Wiley & Sons, Inc. Published 2025 by John Wiley & Sons, Inc.

4

Tectonic Affects and Innovative Attitudes due to Changes During Construction of the Foundations, Frame System, Walls, and Roof

Everybody (the owner, contractor, and the main controller) was excited at the beginning of the construction of Monarga House.

> A big truck came with an excavator on it. When we – me and my friend – understood that it was for my site, we got very excited … They changed the instrument on the excavator and made it ready for digging the foundations.

Construction of the foundations, the reinforced concrete frame system, the brick walls, openings, heat insulation, roof and pergolas formed the first phase of construction of the Monarga House. This was also the only phase when most of the works were done one by one following the earlier items. First the foundations were built, then the frame system. Then came the walls with the openings and heat insulation and roof and pergolas could be placed. The later phases of construction were not like this because each item continued almost through the end of construction.

Foundations – Changes, Tectonic Affects, and Innovative Attitudes

> … Contractor asked … to make the application of the project on the site through satellite connection. I was worried. But it was almost all right.

After spotting the site of the building with the help of a satellite connection, the first step of construction of the Monarga House was the excavation of the site. Figures 4.1a and b show that the excavation started with the excavation

Tectonics as a Process in Architecture, First Edition. Yonca Hurol.
© 2025 John Wiley & Sons, Inc. Published 2025 by John Wiley & Sons, Inc.

Figure 4.1 (a) and (b) Excavation. (c) Addition of the waterproof membrane and the protective layers above and under it. (d) and (e) Reinforced concrete foundations. (f) Formwork for the reinforced concrete tie-beams. (g) Completion of the reinforced concrete tie-beams. (h) Water isolation inside the cells. (i) Gravel filling in the cells.

of the foundations. However, if one examines Figure 3.1 in Chapter 3 which demonstrates the preliminary architectural project, it becomes clear that first, the rear portion of the site should have been excavated to reach a horizontal surface and after that the trench for the foundations should have been excavated. The RONAC (Researcher, Owner, Neighbor, Architect, Controller) warned the contractor about this change. However, the contractor told her that the problems which might occur due to this change could be solved later through earth filling. This was a mistake which initiated some important changes in the project.

The depth of the foundations was also a problem. The contractor did not want to dig very deep because the site was rocky. He admitted that he was worried about

Foundations – Changes, Tectonic Affects, and Innovative Attitudes | **115**

the height of the entrance level. This is a common problem which occurs in the construction of many buildings in North Cyprus. The contractors prefer to build slab on ground types of foundations without excavating.

The RONAC was satisfied that the first line of bricks for the walls of the house were laid at 42 cm above the level of the earth. Figures 4.1d–f demonstrate the reinforcement of the lowest level of the foundations, the column reinforcement, and the positioning of the reinforcement of the tie-beams to raise the building above the level of the earth. Places for columns were also marked with the help of a satellite connection. Finally, Figures 4.1g–i, show the completion of the foundation tie-beams, the application of water isolation into the cells of the foundations and filling the cells with gravel.

These changes during the construction of the foundations triggered some other changes later because the rear of the site remained unexcavated in accordance with the preliminary architectural project. This caused the following three changes with respect to the design project of the building:

- It became necessary to create a different level in the garden which is supported by some low stone retaining walls. This happened because the height of the back terraces became more than 1 m from the level of the earth. It was necessary to add balustrades to the L-shaped terrace at the rear because of the height of this terrace. This was needed because of the TRNC (Turkish Republic of Northern Cyprus) city planning regulations to protect people from falling accidentally from high terraces or balconies. If the rear of the site had been excavated before the foundations, this problem would not have arisen. The RONAC did not want to have balustrades on the terraces and it became necessary to add some stone retaining walls around the rear terrace to create a second level.
- This mistake also resulted in a superfluous stone wall separating the front and back of the garden. This is the wall which divides the site into front and rear parts in the preliminary design project. The wall was superfluous because it was no longer a retaining wall with at least 50 cm of difference in levels on both sides. The level of earth on both sides of this wall became almost the same. In the preliminary architectural project there is a 50-cm level difference between both sides of this wall.
- Again, because of not digging the rear of the site as intended in the project, the ramps which existed in the project to connect the front and rear of the garden, were not built. This continuous stone wall splitting the garden into two meant that a wheelbarrow could not be used for gardening. The RONAC later designed a portable timber ramp to solve this problem.

Another change during the construction of foundations affected the building's sewage system. Since an opening had not been bored in the tie-beams of the

Tectonic Affects due to Changes During the Construction of the Foundations

foundations for the toilet pipe, it became necessary to have the toilet manhole visible at the west of the house because the toilet pipe had to be at a higher level. This affected the aesthetics of the west façade.

Tectonic Affects due to Changes During the Construction of the Foundations

Not all changes cause tectonic affects. Foundations are actually a hidden element of construction. After the building is complete nobody can see the foundations. There is always suspicion about the construction of foundations because of their hidden character. Many people think that corrupt contractors do not dig deep enough and steal reinforcements, and so on.

> *Reflection(s) from Theory*
> Many theories of architecture are suspicious about **hidden elements and components** in buildings. Karl Bötticher and Eduard Sekler's theories of tectonics can be mentioned as one example. Bötticher thought that the structure should realize the representative role of buildings and because of this it should be perceivable.[1] Eduard Sekler preferred building elements and components to be perceivable and not illusory.[2] However, neither of these theoreticians has written anything about the foundations, which are always under the ground and not perceivable.

Five changes occurred during the construction of foundations and three of these initiated other changes which caused tectonic affects. There were no eliminated changes during the construction of the foundations. The stone walls on the site and a second level with stone steps because of mistakes in excavation, caused some pleasing tectonic affects due to the use of stone, which is a natural material. These included the hapticity of stone, and having some transitory layers of different heights around the building.

---

1 Karl Gottlieb Wilhelm Bötticher, "The Principles of the Hellenic and Germanic Ways of Building with Regard to their Application to Our Present Way of Building." In *What Style Should We Build?* Eds: J. Bloomfield, K. Forster and T. Reese. (The Getty Center Publication Programs, 1992[1828]): 147–168.

2 Eduard Sekler, "Structure, Construction, Tectonics." In *Structure in Art and Science*, Ed: G. Kepes. (Wordpress, 1965): 89–95. accessed August 18, 2018. https://610f13.files.wordpress.com/2013/10/sekler_structure-construction-tectonics.pdf.

Foundations – Changes, Tectonic Affects, and Innovative Attitudes | **117**

> *Reflection(s) from Theory*
>
> The painter Francis Bacon was quick at making draft paintings and he ended up with errors which were later regarded as gifts and articulated by the painter as contributing to the affects caused by these paintings. These articulations resulted in impressive details in his paintings.[3] Could making **errors to articulate** them later during the construction of buildings be an artistic approach to the construction process?

Having a stone wall separating the front and back parts of the garden without any difference in level difference between its two sides, is not perceived by many people. However, if it is consciously perceived, it causes negative tectonic affects due to being unreasonable. However, the use of a portable timber ramp for gardening because the ramps listed in the project were not built can be perceived only while gardening.

> *Reflection(s) from Theory*
>
> Since the problem is different for having an extra level around the back terrace which has some positive tectonic affects, this change deserves some different theoretical reflections because it has no positive tectonic affects. The professional project is ignored, and this is a reminder of the **contradiction between the modern professional approach versus the traditional approach** which does not require any scientific approach. This traditional approach is quite common among construction workers in North Cyprus. Many of them prefer to ignore projects and professionals. **Ignorance** is one way of dominating others and the application of power.[4]
>
> There can also be reflections from various theories of ethics. Since both theories of tectonics and theories of architecture contain many connotations about ethics, such as being honest and having building elements and components perceivable (which can be related to virtue ethics because **competence** is addressed),[5] such reflections are reasonable. According to John Rawls, principles of justice are based on **social contracts**. The action of putting a signature onto a contract between a contractor and an client is also defined within such

(Continued)

---

3 David Sylvester, *Interviews with Francis Bacon.* (Thames & Hudson, 2016).
4 Bernard Williams "Philosophy and the Understanding of Ignorance" *Diogenes* 43, no. 169, 1995: 23–36.
5 *Aristotle Nicomachean Ethics.* (University of Chicago Press, 2012[350 bce]).

> *(Continued)*
>
> recognized actions of social contracts. If one signs a contract to do something specific first and then it is not done, this is not ethical.[6] Immanuel Kant's deontological ethics (duty ethics) can also be a source of reflection. Kant suggests that **following rules and laws,** even if we do not like them or we do not support them, is ethical.[7] Building activity also has some rules. Following architectural application projects as much as possible during construction of the buildings is one of these. It can be said that not following the project specifications, and ending up with problems, goes against duty ethics.
>
> There can be reflections from the theories about errors in art and architecture. The error concerning the excavation of the site of the Monarga House is not comparable to the artistic errors. However, it also does not match the concept of **benign errors** (friendly or sympathetic errors) of the architect Alvar Aalto, which are signs of a critical approach towards perfectionist design and construction. Aalto created some irregularities in his Villa Mairea, such as the way the staircase was hung and different types of wrapping of irregularly spaced columns. The marble cover on Aalto's Finlandia Hall is another example of benign errors. The marble on the surface of this building became slightly curved due to extreme climatic conditions. Such a detail can be called a "delayed detail" as well. It is mainly Alvar Aalto, Göran Schildt, and Juhani Pallasmaa who addressed the benign errors.[8] John Ruskin also defended this long ago declaring that "Imperfection is the sign of life ... Irregularities and deficiencies are not only signs of life, they are also sources of beauty."[9]
>
> According to the fragile architecture approach, which also defends **imperfection,** "Clarity of image usually contains hidden repression."[10] This approach is a reminder of the concept of "weak thought."[11] Ushida and Findlay's Truss

---

6 John Rawls, *A Theory of Justice.* 2nd Edition. (Harvard University Press, 1999[1971]).

7 Immanuel Kant, *The Metaphysics of Morals.* (Cambridge University Press, 1996[1797]).

8 Juhani Pallasmaa, *The Eyes of the Skin – Architecture and the Senses.* (John Wiley, 2005). Alvar Aalto, "Inhimillinen Virhe (The Human Error)." In *Nain Puhui Alvar Aalto (Thus Spoke Alvar Aalto)*, Ed: G. Schildt. (Otava, 1997): 282. Alvar Aalto and Göran Schildt, ""Speech at the Centennial Celebration of the Faculty of Architecture" on May 12, 1972 at Helsinki University of Technology." In *Alvar Aalto in his Own Words.* (Rizzoli, 1998).

9 Gary Coates, *Erik Asmussen – Architect.* (Byggforlaget, 1997): 230.

10 Juhani Pallasmaa "Hapticity and Time-Notes on Fragile Architecture" *Architectural Review* 207, 2000: 78–84. accessed September 3, 2020. https://pdfs.semanticscholar.org/e633/c06ae14c8fb9eeaadad27cde25432ac931ac.pdf.

11 Gevork Hartoonian, *Ontology of Construction – On Nihilism of Technology and Theories of Modern Architecture.* (Cambridge University Press, 1994).

> Wall House is an example of fragile architecture. There is continuity between the human body, human movement, and the dynamic flow of the plastic spaces within this building and they cause imperfection in geometry and details. Although the building looks like an adobe building, it was formed by using bent steel elements which were themselves covered with slender steel elements and steel mesh and finally with a shotcrete layer. The building does not have a typical structural system, but it is acceptable.
>
> Marcel Proust also mentions some other types of errors, which cause haptic affects and recall memories. In the last volume of *In Search of Lost Time* (*Time Regained*), he gave several examples of **errors causing haptic affects** to recall past (especially childhood) memories as flashbacks. These haptic affects can be exemplified with broken or imperfect marble on the floor which may cause one to slip. The character in Proust's novel slips again many years later due to the same broken marble and his memories of his lost mother return as a flashback. Proust gave importance to these flashbacks, because they remind people of their lost childhood opportunities.[12]
>
> The problem wall which divides the site of the Monarga House, is as negative as Marcel Proust's errors, which cause haptic affects because there is nothing tectonically positive about it. The other error types can cause positive even artistic tectonic affects. However, the problem wall is also different from Proust's errors because it can only be experienced mentally and cannot cause any haptic affect. If one does not think about the problem with this wall after perceiving it, they may not recognize or experience the error. As a result, it may not trigger any memories. There is nothing positive about it.

Similar to the wall dividing the site of the Monarga House in two, the change in the height of the toilet manhole also caused some negative tectonic affects. Table 4.1 presents the tectonic affects due to changes during the construction of the foundations.

The tectonics of the topography is very much influenced by the changes in the excavation and foundations of the Monarga House.

---

12 Marcel Proust, *Search of Lost Time (À la Recherche du Temps Perdu)*. 7 volumes. (Everyman's Library, 2001[1913–1927]). The titles of the seven volumes are *Swann's Way, In the Shadow of Young Girls in Flower, The Guermantes Way, Sodom and Gomorrah, The Prisoner, The Fugitive,* and *Time Regained.*

120 | *4 Tectonic Affects and Innovative Attitudes due to Changes During Construction*

Table 4.1 Tectonic affects due to changes during the construction of the foundations.

The change in 2	The symbol	The tectonic affect(s)	Responsible person/ contributor	Related changes	Innovative attitudes
Having stone walls around the back terraces because of unintentional excavation.		Use of natural materials, hapticity of stone, having transitory layers with different height levels around the building. These layers provide better continuity with the environment. There is also the tectonic use of topography.	Contractor, the RONAC	Many	No
Having a stone wall with no height difference on both sides of it.		If this is perceived consciously, it can cause negative tectonic affects due to being unreasonable.	Contractor	Many	No
Not making a hole at the tie-beam level to let a pipe reach the toilet manhole.		Resulted in bad detailing and bad tectonics.	Contractor	Many	A possibility but not realized

Eliminated changes/problems are shaded in light gray in the table. Changes which cause another change or originate from another change are shaded with darker gray. Changes that did not cause any other changes are not shaded. Symbols reflect not only the type of change, but also the associated feelings. Lighter colors in symbols reflect positive feelings.

> *Reflection(s) from Theory*
>
> The concepts of *topos* (the site), *typos*, and tectonics are studied by Kenneth Frampton. His concept of *topos* has strong connections with culture.[13] Yi-Fu Tuan suggested the concept of **topophilia**; *topos* meaning the place and *philia* meaning love to express the connections between the material world and humans.[14]

---

13 Brian Kenneth Frampton, *Studies in Tectonic Culture – The Poetics of Construction in Nineteenth and Twentieth Century Architecture*. (MIT Press, 1995).
14 Yi-Fu Tuan, *Topophilia: A Study of Environmental Perceptions, Attitudes and Values*. (Columbia University Press, 1990).

Innovative Attitudes During the Construction of the Foundations

Not boring a hole in one of the tie-beams of the foundation to pass the toilet pipe through to the toilet manhole, which is outside the building, could have provoked an innovative attitude to avoid the negative tectonic affects due to having a high toilet manhole near the back façade of the building. However, this has not happened.

Reflection(s) from Theory

As aforementioned, Immanuel Kant defined ethical behavior as **following rules** whether we like them or not. However, both Thomas Hobbes and John Locke wrote that rules are not for controlling people, but exist to facilitate life.[15] Kant wrote that societies have several weaknesses, and therefore, we cannot expect people to follow rules willingly. On the other hand, Alain Badiou suggested that being open for change is ethical. There is a tension between the ethics which suggest following rules and the ethics which suggest **being open for change**. The philosopher Gaston Bachelard's explanations about the Prometheus complex in his book *Psychoanalysis of Fire* discusses the relationship between following rules and looking forward to changes towards improvement. Bachelard wrote that although families prevent their children from lighting fires to avoid having their houses burnt, all healthy children burn fire. Children aim to burn fire without burning the house. Bachelard's message is if one can do something better than the suggestion of the rule, then that is acceptable. This reveals an important characteristic about human nature which is seeking the betterment of everything including the rules.[16] **Innovations** in building technology are also above the rules, because they are contrary to the old knowledge about technology, but they provide the betterment of those technologies.[17] Thomas Kuhn's concepts of evolutionary and revolutionary changes in science[18] can also be reflected in the concept of innovation as **evolutionary innovations** and **revolutionary innovations**. But this situation in the Monarga House is a **lost possibility of an evolutionary innovation**.

---

15 Thomas Hobbes, *Leviathan – The Matter, Form and Power of Common-wealth Ecclesiastical and Civil.* (Oxford University Press, 2012[1651]). John Locke, *Second Treatise of Government.* (Hackett, 1980[1689]).

16 Gaston Bachelard, *The Psychoanalysis of Fire*, Trans. A. C. M. Ross. 2nd Edition. (Beacon Press, 1987[1938]).

17 Yonca Hurol, *The Tectonics of Structural Systems – An Architectural Approach.* (Routledge, 2016).

18 Thomas Kuhn, *The Structure of Scientific Revolutions.* (University of Chicago Press, 1962).

Frame System – Changes, Tectonic Affects, and Innovative Attitudes

Construction of the frame system started with the reinforcement of the columns. However, although the height of the columns was 3.75 m in the project specification, the height of reinforcement was 3 m.

> Then one day I saw that the reinforcement of columns was installed. I was very happy. But later the main controller called me to say that they were too short. They made them 3.00 meters tall instead of 3.75 meters. I wanted the house to be high to make it less hot during summer. I was shocked. The controller said, "They didn't even make it 3.06. That is common in construction. And this tells me that he is trying to eliminate ending up with "fire" in steel."

This reinforcement had to be changed and then the formwork for the columns was installed and the concrete poured into it. Figures 4.2a–c show the construction process of the columns of the building while Figures 4.2d–f depict the formwork and reinforcement of the reinforced concrete beams and the slab.

> Both the main controller and contractor found the reinforcement of the beams of the application project very insufficient. They also told me that the project might not have been designed for the addition of a second floor later. The controller, who is an experienced civil engineer, made the calculations again and I accepted an increase in the reinforcement of the beams. I recorded these new calculations to my computer. I feel safer now. Then I contacted the chief architect of the firm, who managed the application project, to give me a document which shows that the structure of the building has been designed for two floors.

The RONAC later received that document from the TRNC Chamber of Civil Engineers which allowed the addition of a second floor sometime in the future.

The column reinforcement extends outside to enable the addition a second floor to the building if it becomes necessary. Figure 4.2f also shows the electrical wiring which is threaded through the reinforced concrete surfaces. Figure 4.2g depicts the completed reinforced concrete frame. During this stage the building team twice took test specimens from the concrete to check its quality and to be sure that it is the right type of concrete as specified in the civil engineering project.

Frame System – Changes, Tectonic Affects, and Innovative Attitudes | **123**

Figure 4.2 (a) Reinforcement of columns. (b) Formwork of columns. (c) Removal of column formwork. (d) Formwork for the slab. (e) Reinforcement of the slab and beams. (f) Concrete for the slab is poured in. (g) Reinforced concrete frame. (h) Reinforced concrete slab and beams. (i) Application of the preparatory layer of plaster.

Changes During the Construction of the Reinforced Concrete Frame System[19]

- The reinforcement for the columns was placed 3 m high instead of 3.75 m. This would have had an enormous impact on the space quality as well as

---

[19] Hashtags (#) indicate mistakes which could have been identified and solved by experienced foremen.

the climatic comfort of the building. This change was not accepted, and new reinforcement was installed taking account a 3.75 m ceiling height plus the extension of reinforcement which would enable the construction of an upper floor later on. The RONAC asked the contractor to change the short column reinforcement and did not want extension bars. The contractor had to accept this request (#1).

- The frame of the building was not built sensitively enough for a 3 cm heat insulation layer on its reinforced concrete elements (#2). Heat insulation was a part of the architectural project. This problem could not be solved because of another related problem. The RONAC wanted traditional *leaf plaster*[20] and application of this heavier plaster type on the heat insulation layer is a problem. The building team canceled the heat insulation for the reinforced concrete elements. Since the heat insulation was included in the cost, the contractor compensated for this problem by providing timber frames for the interior doors. The experts who work on climatic problems in Cyprus also said that removal of the heat insulation layer will not be very effective if the walls are built with eco-bricks because the inside and outside temperatures are not radically different in Cyprus. Therefore, this problem ended up having positive tectonic affects due to the presence of timber frames for the interior doors and the application of leaf plaster.
- The main controller (a civil engineer) and the contractor found the reinforcement of the floor beams to be insufficient. The beams were recalculated, additional reinforcement was added, and the remainder was in accordance with the application project.

Tectonic Affects due to Changes During the Construction of the Reinforced Concrete Frame System

There were three changes at this stage of construction. Only one of these changes could have affected the architecture of the building and caused tectonic affects. This change was initiated by the contractor. This is the decrease in the column height which was identified on time and eliminated. If the mistake in the column height had not been identified and stopped, there could have been a serious change in the tectonic and architectural character of the building. This could have changed the scale and proportions of the building. It could have made the building more concealed within the trees. It could also have terminated the civil engineering project, which was only half realized because of the change in the reinforcement of the beams. The other changes during the construction of the reinforced concrete frame system are not perceivable (Table 4.2).

---

20 *Leaf plaster* is "yaprak sıva" in Turkish.

Frame System – Changes, Tectonic Affects, and Innovative Attitudes | **125**

Table 4.2 Tectonic affects due to changes during the construction of the reinforced concrete frame system.

The change in 3	The symbol	The tectonic affect(s)	Responsible person/ contributor	Related changes	Innovative attitudes
Elimination of the change in the column height (rework).		This could have changed the scale and proportions of the building and made the building more concealed within the trees. It contradicts the architectural project, and it causes negative tectonic affects.	Caused by the contractor, identified by the main controller.	0	No

Eliminated changes/problems are shaded in light gray in the table. Changes which cause another change or originate from another change are shaded with dark gray. Changes that did not cause any other changes are unshaded. Symbols reflect not only the type of change, but also the associated feelings. Lighter colors in symbols reflect positive feelings.

Reflection(s) from Theory

Martin Heidegger's critique of humanity's approach to technology is still relevant. He explained the **instrumental approach** to the world as an orientation originating from the scientific approach to achieve economic benefit, and he also wrote that this happens by "enframing" objects and seeing them as "standing reserve." **Enframing** is about defining things in specific ways (towards achieving benefit) so that everybody understands them like that. Seeing forests as sources of wood and ignoring the fact that the trees are alive and there are many animals living in forests, is a good example of enframing. **Standing reserve** is one step further than enframing suggesting that things in nature are there for the benefit of humans and we can go and take them from there whenever we wish. Jacques Ellul's theories also support Heidegger's critiques.[21] John Habraken wrote that such an approach to economy also exists in the building sector, and it causes **monotonous** and

(Continued)

---

21 Martin Heidegger, *"The Question Concerning Technology" In The Question Concerning Technology and Other Essays*, Trans. William Lovitt. (Garland Publishing, 1977[1954]a): 3–35. Martin Heidegger, "The Age of World Picture." In *The Question Concerning Technology and Other Essays*, Trans. William Lovitt. (Garland Publishing, 1977[1954]b): 115–154. Martin

> **(Continued)**
>
> **meaningless** environments. He wrote that *supplier-driven models* cause this situation rather than *user-driven models.*[22] Radically decreasing the column height of a building, even though the project suggests otherwise, can be explained with the concepts of the instrumental approach, enframing of building elements and components and causing meaninglessness. If the same error is repeated by many contractors within the same environment, then this also causes monotony.

Innovative attitudes during the construction of the frame system:

- Leaving an ambiguous relationship between the heat insulation layer on structural elements and the leaf plaster during the application project phase affected the construction of the frame system too. Not building the frame system carefully to have 3 cm of space for the application of the heat insulation layer on the reinforced concrete elements led to an opportunity for innovation. There was only 1 cm difference between the surfaces of the reinforced concrete elements and the brick wall surfaces. The main controller said that during the plastering stage the problem could be solved. However, there was also a second problem related to the application of the heat insulation. This problem was about the weakness of the heat insulation layer in carrying the leaf plaster. Finally, the heat insulation of the reinforced concrete elements was not done because this change was supported by an academic who specializes in heating and cooling problems of buildings.

> *Reflection(s) from Theory*
>
> As previously asked: Could ambiguity in architectural projects be suggested as a reason for innovation? It is worth revisiting Mahairi McVicar's explanations regarding **ambiguity** in application projects.[23]

Heidegger, *"Science and Reflection" In The Question Concerning Technology and Other Essays,* Trans. William Lovitt. (Garland Publishing, 1977[1954]c): 155–182. Jacques Ellul, *The Technological Society,* Trans. J. Wilkinson. (Alfred A. Knopf, Inc. and Random House Inc., 1964). Jacques Ellul, *What I Believe,* Trans. G. W. Bromiley. (Eerdmans, 1989).

22 John Habraken, *The Structure of the Ordinary.* (MIT Press, 2000).

23 Mahairi McVicar, *Precision in Architecture – Certainty, Ambiguity and Deviation.* (Routledge, 2019).

Walls, Heat Insulation, Openings, Changes, Tectonic Affects, and Innovative Attitudes

After completion of the frame system, construction of the brick walls started. There were two types of bricks. The inside walls were built with ordinary bricks and the outside walls were built with eco-bricks to provide better heat insulation. Figures 4.3a–c show the construction of the brick walls, with two types of bricks interlocking and the use of steel profiles as the lintels of small openings. Figures 4.3d–f depict the reinforced concrete tie-beams which were connected to the side surfaces aided by the injection of epoxy into these surfaces for reinforcement. The project contained reinforced concrete lintels for all openings. There were several changes at this stage of construction due to the preliminary and application projects.

Figure 4.3 (a) Water isolation under the brick wall. (b) Interlocking two types of brick walls. (c) Steel profiles as lintels for small openings. (d) Installation of steel bars to connect the reinforced concrete lintels of longer openings to the structure. (e) Reinforcement for longer openings. (f) Formwork for the lintels of longer openings.

... Longer lintels have been built with reinforced concrete. They drilled holes into the columns and put reinforcement in. This reinforcement extended 80 cm out. They put epoxy into the holes, added reinforcement, and then poured concrete into the formwork. They kept the formwork for 10 days. I was worried about this application. Because I thought the longer lintels were also going to be steel. However, the main controller told me that this is a common application.

Changes while Building the Brick Walls and Making the Openings in Them[19]

- The RONAC agreed to having the interior walls constructed with normal red bricks, because no heat insulation problems were expected within the rooms of the house.
- The house was designed to have typical reinforced concrete lintels over the openings. However, this was changed during the construction of steel L profile lintels for small openings and reinforced concrete lintels, which were supported by steel bars injected into reinforced concrete surfaces on two sides of the openings, for the longer openings.
- The heights of the lintels of the openings were not determined according to the preliminary/application projects (#3). This caused a difference in height between the interior doors and the openings on the exterior walls and also affected the height of the pergolas. Since nobody was aware of this at that time, the problem appeared later. This change caused negative tectonic affects by creating a chaotic situation inside the house.
- The height of windows from the floor was high in the preliminary and application projects. A foreman warned about the possibility of bad ventilation through these windows in the future (#4). This was accepted by the RONAC and the heights of most of the windows from the floor level were lowered. This was contributed by a foreman which affected the function of all the windows.
- Heat insulation on the reinforced concrete frame elements became a problem and there were several changes of decision about this issue.
  - Bricks were not extending outside, because the formwork of beams and columns had expanded slightly (1.5 or 2 cm) and this created the problem of the bricks extending only 1 or 1.5 cm which is not sufficient for the placement of heat insulation on the frame elements. Finally, the insulation layer on the structural elements was canceled.
  - The suggested insulation materials cannot hold leaf plaster. It is an architectural design mistake to expect leaf plaster to be applied onto heat insulation. There were two mistakes: one was the contractor's mistake (about precision) and the other was the RONAC's mistake (about imagining a strong mesh to hold leaf plaster).

- There was also the idea of changing the plaster type and using a specific type of plaster with insulation properties. However, an experienced architect warned about the problem of applying two different types of plaster on a surface which causes cracks.
- Finally, the insulation of the frame elements was canceled. Because of this, the contractor accepted using timber frames for the interior doors and making the area of leaf plaster larger. Plastering was done almost exactly as planned in the preliminary design project and not as it was changed during the application project to decrease the cost. Finally, something very close to the suggestions of the preliminary project had been achieved.

Reflection(s) from Theory

The way the lintels were redesigned and built during construction (the use of steel for shorter ones and applying steel injection for the bars in the longer reinforced concrete ones) recalls a reflection from Gevork Hartoonian who wrote that although honesty is important for modern architecture, reinforcement in the concrete elements and their details have a hidden character. These **hidden details** cannot be subjected to joint-disjoint (*montage*).[24] These are details in which *logos of techne* (technical dimension in techne) is dominant and *techne of logos* (artistic dimension in techne) does not exist.[25] Such details are different from the details which can cause tectonic affects.

Does having different heights for all openings (windows and doors, and later the heights of wardrobes and cupboards, will also be included to this situation) cause **imprecision as a design quality**[26] which is contrary to precision? If precision is exaggerated as in the case of Hermine's house which was designed by the philosopher Ludwig Wittgenstein, it becomes like a "dwelling for the gods."[27] However, imprecision can also be an outcome of artistic design like the Truss Wall House of Ushida and Findlay (which was also cited as an example of fragile architecture). However, it is more appropriate to categorize the change of the window and door heights in the Monarga House as an ethical problem about **not following the professional rules and agreements**. It can also be seen to result from **ignorance** which is a way of dominating others.

---

24 Hartoonian, *Ontology of Construction,* Chapter 1

25 Marco Frascari, "The Tell-the-tale Detail." In *The Building of Architecture.* (1984). accessed August 19, 2018. https://uwaterloo.ca/rome-program/sites/ca.rome-program/files/uploads/files/frascari-m-the-tell-the-tale-detail-3-a.pdf.

26 Yonca Hurol, *Tectonic Affects in Contemporary Architecture.* (Cambridge Scholars Publishing, 2022): 175–180.

27 Stuart Jeffries, "A Dwelling for the Gods" *The Guardian*, January 5, 2002, accessed February 19, 2025. https://www.theguardian.com/books/2002/jan/05/arts.highereducation.

Tectonic Affects due to Changes During the Construction of Walls, Heat Insulation, and Openings

There were five changes which happened during the construction of walls, heat insulation layers, and openings. Only two of these can cause some tectonic affects. The heights of the windows were not consistent with the architectural project. Because of this, the heights of openings, doors, and wardrobes and cupboards were different, and this created different height levels in the house and negative tectonic affects.

The contradictory relationship between heat insulation on the reinforced concrete elements and the type of plaster could have caused changes in tectonic affects. However, since the heat insulation layer was canceled and leaf plaster was applied, the negative change in tectonic affects had been avoided. If the leaf plaster was not done, this could have had a considerable impact on the tectonic qualities of the building. This was one of the problems which triggered many other changes during the construction process.

The column labeled "innovative attitudes" in Table 4.3 illustrates the complexity of the relationship between innovative attitudes and tectonic affects. This column should be interpreted in conjunction with the symbol indicating whether the tectonic affect is positive or negative. Although both the first and third rows represent unrealized innovative attempts, one results in positive tectonic affects while the other results in negative ones. This discrepancy arises because the goal of innovative attitudes may differ from achieving tectonic affects. For example, the unrealized innovative possibility of applying heat insulation to structural elements led to positive tectonic affects, as it allowed for the use of leaf plaster, a haptic tectonic detail. However, exploring the relationship between tectonic affects and innovative attitudes is beyond the scope of this book.

> *Reflection(s) from Theory*
> The hierarchy of values may change from project to project, from person to person, and from culture to culture. However, there can also be reasonable approaches to the **hierarchy of values** such as comfort and aesthetics.[28] Even if one accepts different approaches to the hierarchy of values in buildings, understanding them tells a lot about the decision makers.

---

28 M. Pultar (ed.), *Mimarlık Bilimi: Kavram ve Sorunları*. (Science of Architecture, Concepts and Problems, Çevre ve Mimarlık Bilimleri Derneği, 1978).

Table 4.3 Tectonic affects due to changes during the construction of walls, heat insulation, and openings.

The change in 4	The symbol	The tectonic affect(s)	Responsible person/contributor	Related changes	Innovative attitudes
Heights of windows were not done according to the project.		This causes a negative tectonic affect due to being un-architectural and unreasonable. It causes bad detailing in the whole house.	Caused by the contractor	Many	A possibility, but not realized.
Windows' height from the floor decreased for better ventilation.		This provided better lighting inside the rooms.	Suggested by a builder	0	None
Having heat insulation on structural elements became impossible, but this enabled leaf plaster on walls.		The elimination of heat insulation enabled the use of leaf plaster which creates haptic tectonic affects and causes continuity with the historic buildings in the same vicinity.	Decided by the RONAC	Many	A possibility, but not realized.

Eliminated changes and problems are shaded in light gray in the table. Changes which cause another change or originate from another change are shaded in dark gray. Changes that did not cause any other changes are unshaded. Symbols reflect not only the type of change, but also the associated feelings. Lighter colors in symbols reflect positive feelings.

Innovative Attitudes During the Construction of Brick Walls, Openings, and Heat Insulation

There are two possibilities for innovative attitudes in this process.

- Whatever began as an innovative attitude during the construction of walls, openings, and the application of heat insulation is the same problem that was initiated during the construction of the reinforced concrete frame system, and it is about applying the heat insulation layer and the leaf plaster together. This potential for an innovative attitude has not been realized.
- It may also be possible to see the application of reinforced concrete lintels with the injection of steel bars with epoxy as an innovative attitude in this process. However, since this is not perceivable, it cannot cause any tectonic affects.

Roof, Heat Insulation on the Ceilings and Pergolas – Changes, Tectonic Affects, and Innovative Attitudes

We chose the roof tiles from a good firm. They will not be a shiny tiles, but they might collect dust. I am a bit worried about the color of them. We have to choose the colors of the shutters and fascia on the eaves accordingly.

The construction of the roof gave rise to several types of problems. It can even be said that everything was a problem. Figures 4.4a–c show the construction of the timber skeleton of the roof with triangular windows. Each timber element was isolated against water. Figures 4.4d–f show the timber cover on the roof structure. The cover on the eaves was different from the cover on the parts which will not be visible later. The timber surface was covered with water isolation sheets and linear timber elements were placed to hold the tiles. Figures 4.4g–i show some changes in the details of the roof as well as the placement of the ridge tiles. These changes are explained below. Figures 4.4j–l show the stages of construction of the pergolas.

Roof, Heat Insulation on the Ceilings and Pergolas – Changes | 133

Figure 4.4 (a) Construction of the timber roof structure. (b) Construction of triangular roof windows. (c) Purlins isolated against water with liquid material. (d) Different timber covers for eaves. (e) Timber cover on the purlins protected by water isolation sheets. (f) Linear timber elements which will support the roof tiles incorrectly placed horizontally. (g) Horizontal timber elements removed and placed over vertical linear elements. (h) A roof detailing problem which eliminates water in the roof from escaping solved by boring small holes for water egress at the edges of the eaves. (i) Positioning of roof and ridge tiles and since the ridge tiles were too high, concrete was applied on two sides of them. (j) The installation of the timber structure of the pergola. (k) Development of structures of that pergola. (l) Completed pergolas with an angle problem in the right-hand pergola.

Changes During the Construction of the Roof, Heat Insulation, and Pergolas[19]

There were many problems and changes during the construction of the roof, which was planned to be a ventilated cool roof, and the attic was also designed to contain a 1-ton water tank and a boiler.

- The angle of the timber roof structure was wrong at the beginning (#5). This made the roof much lower than it was in the original project. Since this was going to make a functional (the roof contains a tank and a boiler in the project) and aesthetic (this totally changes the image of the building) impact on the building, the RONAC did not accept this change.
- Some timber elements in the roof structure were damaged while changing the slope. This happened because the builder continued to use the same elements. He added elements to other ones, and he also used some damaged ones. The main controller did not spot them and the RONAC, who could not climb up to the roof, had to maintain them later.
- The hardboard cover of the roof frame was made with bits and pieces in some places and without properly connecting them to the frame. These pieces fell out later, and the RONAC had to fasten them to their positions again. There are also some odd and ugly opening details (of the triangular windows) in the attic as well.
- The triangular windows on the roof were placed much lower than in the original project at the beginning (#6). This was changed. Since the main controller suggested not having any openings at the top and the bottom of the roof for ventilation due to the builder's lack of experience, it was thought that putting the triangular windows in the middle would ensure better ventilation. However, the roof was not finalized as a good cool roof which is ventilated and isolated.
- The triangular windows were built smaller than they were in the project (#7). This affected the effectiveness of the roof as well as its aesthetics.
- The RONAC decided to put timber around the triangular windows in the roof and this increased the haptic tectonic quality of the building.
- The sills of the triangular windows on the roof were not built correctly. The worker thought that the aluminum frames would be placed inside the opening. But then the ventilating area was getting very small. The RONAC asked the aluminum firm to put their frames outside. However, then, the sills should also have been outside. A foreman changed them (#8).
- The dark green fascia of the roof was repainted in the natural color of timber.
- The timber cover under the eaves of the roofs and pergolas was changed by the contractor during the tendering process. The contractor suggested using a good quality material and this improved the aesthetic quality of the eaves and

Roof, Heat Insulation on the Ceilings and Pergolas – Changes | **135**

pergolas. This also gave way to having more timber in the house. The connection of the timber posts to the floor was not correct in the application project either. This was also changed by the contractor during the tendering process.

- The height of the chimney was under the roof level. This was going to cause smoke problems for the fireplace. The chimney height was extended by making it 60 cm above the highest roof level as it was in the preliminary project. This also made the chimney visible.
- The inner roof detail was applied wrongly by eliminating the thin vertical timber elements under the horizontal ones which are used to attach the roof tiles. This wrong detail could retain water in if any came in for some reason. The application project was different, so this detail was changed.
- Water egress from the edges of the roof was not possible because of the wrong detailing. Water isolation did not cover the top of the edge beam. The builder used insulation liquid which is used in swimming pools for that detail. He also had to bore holes with a drill at the edges of the eaves to let the water escape if it comes in. However, this is not a good solution because the holes may be filled in later, and they might also cause deterioration of the timber.
- There was also a problem with the ridge insulation detail. In the suggested details, the insulation sheets continue underneath the ridge and extend over the edge beam. This eliminates water coming in from the ridge and it also lets the water out from the edge. However, they put the secondary elements over the corner roof beams without putting insulation between the beam and the secondary element. The purpose of the secondary element is to nail the ridge tiles. The roof ridge tiles were too high because of the wrong detailing. The builder had to cover both sides with concrete (#9). Later the builder wanted to paint these concrete parts red, but the RONAC preferred to keep them in their natural color which is gray. Some people like this, and some do not. The RONAC likes it.
- The roof heat insulation material on the ceiling was changed to increase the quality. But then it was cut because the overflow pipe for the tank in the attic was the wrong size. Roof details including the insulation detail were problematic. Having a cool roof with a tank and a boiler inside was too ambitious for North Cyprus conditions. There was also too much ambiguity in these decisions.
- The problem about protecting the outdoor timber elements could not be solved. Varnish was used on the fascia of the roof, and it was applied very late. This caused deterioration in the timber. The RONAC wanted to have a better quality of timber protection at the beginning because the building is close to the sea. Also, once applied, varnish cannot be replaced. The RONAC planned to add small pieces of timber to cover the rotten parts, but this also never happened. A carpenter suggested burning the varnish off, but the RONAC

found that solution unacceptable. This caused tectonic problems because all the timber elements looked a bit worse because of wear. A carpenter added some yellow paint to the varnish, and this made the fascia look better.

- The issue of having stairs or a ladder to the attic has changed several times. Having an automatically opening staircase from the attic into the house was canceled because that type of staircase is not appropriate for such high ceilings. The RONAC designed several solutions for this but since the entrance to the attic is very close to the fireplace none of them were applicable. The RONAC thought about having an aluminum trap door to the attic but later she decided to have a timber one. "Then the carpenter showed me the timber trap door to the attic. I was shocked. He made that at home. It was nice and painted with a protective material. It was very simple." This timber trap door contributed to the presence of timber tectonics in the house. The RONAC has a portable ladder to reach the attic.

- The RONAC identified a problem when she was at the house with one of her architect friends. They found that when it rains, there is a problem with the front door area because of water coming from one of the diagonal roof hip lines. They left the front door open for a short time and water even came into the house and soaked the wardrobe doors. Two diagonal roof hip lines on the roof had this problem. The problem has been solved by adding two rain chains under those hip lines. "I asked the builder about the water coming from the roof at the entrance area. He suggested putting a thick chain detail there and also to control the water with a solid surface where it falls. This needs a detail." These rain chains formed tectonic details which were designed after the construction of the house. The photograph in Figure 3.2d in Chapter 3 shows one of these rain chains.

- When the RONAC and her friend were at home, they also saw how water moves around the house when it rains heavily. The area beside the front door becomes a pool. The drainage conduits are not sufficient. Later, that was solved through sloping the surfaces and stabilizing the earth.

- A carpenter said that the timber used outside the house was not redwood as specified in the application project (#10). It was yellow pine and not good. This carpenter even suggested changing the corner posts of the pergolas. The RONAC did not accept that.

- The solution to the problem of the timber posts has changed several times. The timber posts under the pergolas were made from of two pieces. It was not specified in the original project. The RONAC and the main controller preferred having metal belts around them to avoid them separating from each other. However, they canceled these metal belts later. The RONAC has to keep her eyes on these posts. "Yesterday I saw on the site that one of the posts of the main pergola (the one at the corner) sits on a piece of marble which is not supported by anything. I am worried about that pergola now."

- The distance between the level of the roof and the pergolas was larger in the preliminary project. Because of the high windows this distance was shorter (#11). "That day I was worried about the distance between the main roof and the top of the pergola. We found out that the window height has been changed and this had consequences for the height of pergolas. We finally fixed it." The RONAC worried about pigeons building nests in those narrow places as a result of this change. She thought about having wire mesh between the roof and the pergolas to prevent them from making nests. However, this was not applied, and there was no problem with pigeons.
- The extension of the pergolas towards the outside was changed to 50 cm rather than 60 cm in order to get more winter sun inside (#12).
- Heat insulation over the reinforced concrete ceiling was also a problem which caused several changes. The roof was designed as a cold roof which has air intake from the bottom of the roof, air exit at the top of the roof and a heat insulation surface on the ceiling plus the triangular windows. Therefore, the attic becomes a relatively cool place and heat in the attic does not negatively affect the spaces underneath. However, in the end, the roof had only the triangular windows to take air in and to provide an exit for it. The small top and bottom holes were not made because the main controller was not sure about the quality of construction. Also, the attic was designed to contain a water tank and a boiler, and they required some precautions for providing water isolation as well. A pool which had water isolation on it was built under the tank and the boiler. However, this pool obstructed the heat insulation on the ceiling. Also, there was an exit pipe to avoid flooding in the attic. Since the thickness of this pipe was very small it had to be changed later, and it became necessary to cut the heat insulation layer on the ceiling. Therefore, the roof cannot be a perfect cold roof.

> *Reflection(s) from Theory*
>
> There are some errors with no positive value amongst the above-mentioned changes. The use of damaged timber elements, adding previously cut pieces to form the timber structure of the roof, making the cover on the frame with bits and pieces of hardboard and cancelation of the top and bottom ventilation of the cool roof can be shown as examples. Before discussing the ethical problems behind these changes, it is worth remembering that these are happening in the **attic**. Gaston Bachelard compared the poetic meanings of attic and basement.[29] In comparison to the basement, the attic is towards the sky; it is

(Continued)

---

29 Gaston Bachelard, *The Poetics of Space*. (Beacon Press, 1994[1958]).

> **(Continued)**
>
> lighter; it is playful. An attic represents the human mind while the basement represents fear of death and the underground world. Having ethical problems in the construction of a roof which represents the human mind sounds strange.
>
> The examples given of the detail errors in the roof are problems with respect to deontological (duty) ethics which is about **following rules**, and the virtue ethics of Aristotle which is about **competence** during production. Aristotle suggested giving the job to the person who knows it the best.[5] The use of damaged elements and bits and pieces of hardboard to cover the roof structure matches serious problems about following professional and artisan rules. Adding previously cut pieces to form the timber structure of the roof also raises the alarm about the competence of the builder as well as his **responsibility** towards his duty. Cancelation of the top and bottom ventilation details of the cool roof suggested by the main controller shows that the controller was aware of the builder's lack of competence. The builder simply cannot do it, and he has no confidence about doing it. However, virtue ethics in modern architectural theory, which advocates for honesty in the use of materials and details, suggests using knowledge appropriately rather than deviating from it to create artistic and tectonic affects.

There was thunder on Saturday night, and I was worried about my house. I was worried about having the roof fly off or having the pergolas collapse. But there was nothing wrong in the morning. I visited the house as usual as I do every day, and saw that there are no problems, and I put the dead bird that I found on the earth in front of the north terrace.

Tectonic Affects due to Changes During the Construction of the Roof

There were 23 changes during the construction of the roof and pergolas. However, only seven of them can cause tectonic affects. It can be observed at this point that most of the changes which occurred during the construction of the roof were not visible changes. Although there are very serious problems with the roof, many of them do not cause any tectonic affects. Is this happening because the builders are more careful about visible problems?

Roof, Heat Insulation on the Ceilings and Pergolas – Changes | **139**

Reflection(s) from Theory

Neil Leach wrote about the concept of camouflage. According to him **camouflage** in architecture indicates a desire to relate to people.[30] However, there can be various reasons for architectural elements and components not to be visible. Certain items might be hidden intentionally. However, for some elements and components it might be natural not to be visible and to be covered. If camouflage is done purposefully and if it causes the intended affects, then it can be accepted that it is done in consideration of better relations with others. However, having an element within the hidden layers of the roof does not match with the concept of camouflage. Still, if an element or component is badly built, knowing that it cannot be perceived later, this idea deserves another concept with some negative meaning. Maybe this situation can be described as **camouflage of noncompetent production**.

Some changes during the construction of the roof could have caused negative tectonic affects, However they were avoided. These are as follows:

- Decreasing the height of the roof due to a change in the angle of the roof elements which also cancels the triangular windows and the presence of a water tank and a boiler in the attic.
- Decreasing the height of the chimney under the level of the roof.

However, there are also some other changes which can cause tectonic affects. These are:

- The timber posts are made of two pieces and some pergola elements have additional pieces. These are visible bad details which can add to the negative tectonic affects caused by the building. Several solutions developed during the construction to solve the problem of these timber posts. However, none of the solutions has been applied.
- The triangular windows are smaller than they were designed. They could have been tectonically more influential if they were larger.
- Surrounding the triangular windows with timber elements, using a higher quality timber surface in the pergolas, keeping the color and texture of timber on the

---

30 Neil Leach, *Camouflage*. (MIT Press, 2006).

Table 4.4 Tectonic affects due to changes during the construction of the roof and pergolas.

The change in 5	The symbol	The tectonic affect(s)	Responsible person/contributor	Related changes	Innovative attitudes
Having timber posts and some other timber elements in pieces.		Negative tectonic affects can be caused by being unreasonable about details.	Problem caused by a builder/contractor.	0	No, new pieces of timber should have been used.
Having smaller triangular windows.		This change decreases the planned tectonic meaning of the roof.	Problem caused by a builder/contractor.	0	No, the technique is the same for both small and large windows.
Having the high ridge tiles covered with concrete from both sides.		This application created gray lines at the corners of the roof. This gives an emphasis to the roof and creates a tectonic detail.	Builder	0	Yes, realized.
Use of timber in many places related to the roof. This started with the change of the material and post connection details in the pergolas. Continued with interior door frames.		Use of a natural material with haptic tectonic affect.	This started with the contractor and was continued by the RONAC.	Many	None

Damaged varnishing of the posts and the fascia because of delays.		This can cause a negative tectonic affect due to being careless and unreasonable.	The problem was caused by the contractor and could not be solved.	0	Unrealized possibility
Adding two rain chains to the building.		Tectonic affect due to detail design.	Suggested by a builder and designed by the RONAC.	Many	Realized innovative attitude
Elimination of the decrease in the height of the roof due to a change in the angle of roof elements. Rework.		This change could have changed the image of the building a lot. It is against the project specification, and it causes negative tectonic affects.	The problem was caused by the builder and contractor, identified and resolved by the RONAC.	0	None

Eliminated changes and problems are shaded in lighter gray in the table. Changes which cause another change or originate from another change are shaded in darker gray. Changes that did not cause any other changes are unshaded. Symbols reflect not only the type of change, but also the associated feelings. Lighter colors in symbols reflect positive feelings.

142 | *4 Tectonic Affects and Innovative Attitudes due to Changes During Construction*

fascia of the roof and having a timber trap door to the attic contributed to the use of timber, which has haptic tectonic affects, in the building.

- The use of varnish instead of quality water protection for timber surfaces and ending up with a shabby impression of many timber elements, such as the fascia and posts, "Today the contractor called me to recheck the color I have chosen for the timber varnish. The contractors were worried about the color of wood putty. Later they said there is no problem … So, the color will not be shiny and it will be transparent."
- The two rain chains were not specified in the architectural project. They were designed and built after the construction to solve water problems caused by the roof. They contribute to the tectonic affects in the garden and also the interior space (Table 4.4).

Reflection(s) from Theory

Just like shortening the columns, decreasing the height of the roof changes the mass of the building has radical effects on the building's image. The major issues of secular contemporary architecture are the **form and image** of the building because some materialistic and poetic issues of traditional architecture have lost their importance. This has increased the importance of the abstract form of buildings.[31] Secularity fueled science, technology, and the productivity of humans. The new materials and systems are associated with secularity especially because of the rise of professions. Economy and the building market became influential on architecture in a different way from the traditional paradigm. This led to the problem of technology which can be expressed by Martin Heidegger's concepts of *enframing* and *standing reserve*. However, secularity does not necessarily mean the loss of meaning. The **meaning** of modern architecture is usually related to its form and image. Such images are sometimes created with the help of original forms (as in the case of parametric design) or they are borrowed from existing images or clichés. If clichés are imitated to create the image, then this is called **sensational image making.**[32] The ceiling height of the Monarga House was inspired by the high internal spaces of traditional Cypriot houses. The roof was also designed to be high to make it more visible because red pitched roofs are known the world over. Red roofs give meaning to houses. However, there is no sensational image in the Monarga House. Therefore, decreasing the height of both columns and roof destroys the form, image, and meaning initiated by

---

31 Francis D.K. Ching, *Architecture: Form, Space and Order*. 4th Edition. (Wiley, 2014).
32 355:1937 Neil Leach, *The Anaesthetics of Architecture*. (MIT Press, 1999).

Roof, Heat Insulation on the Ceilings and Pergolas – Changes | **143**

the architectural project of the Monarga House combining some modern and traditional features.

Initiation of the use of stone and consideration of the quality of timber surfaces under the eaves cause **tectonic affects due to materials**.[33] The **heaviness** and coldness of stone and the **lightness** and warmth of timber can be related to poetics.[34] **Hapticity** of these materials should also be considered because of the rough, cold, and humid surfaces of stone and the texture of water fibers in timber. Juhani Pallasmaa wrote that even the shadows of these materials are different from each other.[35] Using them together causes tectonic meanings which balance each other.

There is also a negative and unrecoverable tectonic affect due to materials in the Monarga House and that is because of the use of varnish for timber instead of a better protective liquid. This causes the fascia of the roof to deteriorate and cause negative tectonic affects.

It is also possible to mention that some changes caused **tectonic affects due to details**.[36] These include timber posts with metal belts, decreasing the distance between the roof eaves and pergolas, and decreasing the length of the eaves of pergolas to get more winter sun inside. The first, which was not realized, could have caused aesthetic results while the latter two cause changes in system details concerning the continuity between **the building and the natural environment**. Although the quality of these details cannot be compared with the surprising tectonic affects caused by the *butterfly beams* in Alvar Aalto's Saynatsalo Town Hall, they still cause tectonic affects.[37] Architectural details are generally seen as functional, and the architectural ornaments are seen as functionless. However, details which can cause tectonic affects combine the functional and functionless characteristics as Theodor W. Adorno suggested to avoid **fetishist characteristics of aesthetic detailing**.[38] This approach goes against the application of **standard repetitive details**.

The roofer's lack of experience resulted in the ridge tiles on the roof being placed too high. He had to cover them with concrete. Since concrete was not painted red, they became visible. These gray borders on the edges of the red roof can be regarded as a **benign error**.

---

33 Hurol, *Tectonic Affects in Contemporary Architecture*: 29–41.

34 Bachelard, *Poetics of Space*.

35 Juhani Pallasmaa, *The Eyes of the Skin – Architecture and the Senses*. (John Wiley, 2005).

36 Hurol, *Tectonic Affects in Contemporary Architecture*: 41–48.

37 An Internet video presents these trusses in three dimensions and describes them using the term butterfly beams. This video also depicts some drawings of these trusses, accessed July 7, 2019. https://www.youtube.com/watch?v=3I6CIV7gC-c.

38 Theodor W. Adorno "Functionalism Today" *Oppositions* 17, 1979: 31–41.

I was worried about the timber elements outside the house because it rains and there is strong sun after rain ... I became very sad because the surfaces of all fascia were slicing.

Innovative attitudes during the construction of the roof and pergolas:

- The problem of having no outlet for water which somehow goes between the layers of the roof triggered an innovative attitude which resulted in a detail which has not been applied before. However, the RONAC believes that boring holes in the timber surface to let water out is not a good solution to this problem. This decision was agreed by all parties including the RONAC, the main controller, and the builder.
- The change in heat insulation material on the ceiling is a result of an innovative attitude. This was contributed by the main controller and was supported by all parties. However, this idea did not fully reach its target because of two problems. The pool area under the water tank and the boiler did not have a heat insulation layer. The change in the small water exit pipe caused a cut in the heat insulation layer. Therefore, this innovative attitude was half successful.
- The details of a cool roof have not been fully applied because of worries about bad workmanship. This caused a change in the placement of the triangular windows. However, this was not the best solution which could have been achieved. This problem did not lead to an innovative attitude. The details of the cool roof do not cause any tectonic affects; however, the placement and dimensions of the triangular windows cause tectonic affects.
- The change of the timber cover material under the eaves of the roof was a result of the innovative attitude of the contractor. He suggested a higher quality material, and this became successful with positive tectonic affects. He also changed the connection of the timber posts to the floor. His knowledge of the building market would have influenced his successful suggestions.
- The ridge tiles of the roof being too high was the builder's mistake. He applied water insulation sheets and covered the sides of the tiles with concrete to solve this problem. This was the innovative attitude of the builder and since these lines of concrete cover on the roof were not painted and perceivable, they cause tectonic affects.
- The rainwater coming from the diagonal hip lines of the roof was eliminated by adding two rain chains to the roof. This was the innovative attitude of a builder, a carpenter, and the RONAC because the top part of these rain chains was timber, and this was not a usual solution. These timber details work well, and they also cause tectonic affects.
- The idea of putting metal belts around the timber posts which were built as two pieces, to stop them separating from each other was a result of the innovative attitude of the main controller. However, because of terminating with the

contractor, this has not been done. If metal belts are applied, they could have caused some tectonic affects.

The builder said that if he had the system detail sheet of the project these detail problems would not have occurred. Since many people do not have details and they ask for the cheapest solution, he started producing that type of stuff. I was telling the contractor that the builder did not know his job. I think like this for that reason. In any case, I know now that most of the roof details in this country are wrong. Even the main controller told me that the details I am asking for can only be seen in the Domus type of journals.

Reflection(s) from Theory

The holes bored under the eaves to let water out, the problematic relationship of the heat insulation layer on the ceiling with the water tank and boiler, and the need for designing two rain chains because of a design mistake can be seen as innovative attitudes with different characteristics. The first two are not fully successful. The second one raises questions about the need for **management of change** as **management of ambiguity**[39] which usually occurs when more than one detail must be integrated. Not all changes occur due to ambiguity but some changes which initiate innovative attitudes, occur because of ambiguity in design. On the other hand, the rain chains, are successful. These rain chains also form tectonic details. They became **errors to articulate**.[40] The effects of roof hip lines were ignored during architectural design, and this created a serious error which had consequences for the use of the house as well as the deterioration of materials and the nearby environment. Articulation of this error with thick chains, stone pots, and stone pieces solved the problem and formed a **tectonic detail**.[36]

Conclusion

This chapter demonstrates that changes occurring in the early phases of construction significantly impact the form and image of the building. Such an impact should be controlled. These changes are as influential as those that occur during the application project and the tendering process. Figure 4.5 presents the changes causing tectonic affects that occur during these phases and the relationships between them. There are two sets of multiple changes within Phase A, relating

---

39 McVicar, *Precision in Architecture – Certainty.*
40 Sylvester, *Interviews with Francis Bacon.*

4 Tectonic Affects and Innovative Attitudes due to Changes During Construction

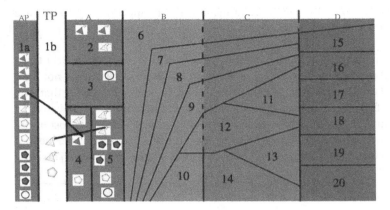

Figure 4.5 The map for the tectonic process for Phase A - Application project (AP, 1a) and tendering process (TP, 1b), construction of foundations (2), the frame system (3), walls and openings (4), the roof (5), plastering and painting (6), the electrical system (7), the mechanical system (8), ceramics (9), water isolation (10), the windows, shutters, and front door (11), the interior doors (12), the wardrobes and cupboards (13), the fireplace and chimney (14), the garden walls (15), drainage (16), the garage and concrete work in the garden (17), metalwork (18), the pathways (19), and timber work in the garden (20).

to the application project (AP) and tendering process (TP) phases. Clearly, these phases caused changes in the construction of walls, openings, heat insulation (4), and roof and pergolas (5). The changes eliminated during this phase of construction indicate a significant reduction in impact on the form and image of the building.

Table 4.5 presents the number of positive and negative tectonic affects and innovative attitudes within Phase A. As highlighted in the table, the high number of changes affecting tectonics during the construction of the roof and pergolas led to additional innovative attitudes.

It should also be noted that there are many more changes at this stage of construction (especially during the construction of the roof and pergolas), but since they are not perceivable, they cannot cause any tectonic affects. This raises the question of whether focus should be on perceivable aspects of construction while neglecting the less visible parts. This chapter also demonstrates that the relationship between tectonic affects and innovative attitudes is complex, as the goal of innovative attitudes may not always be to achieve tectonic affects.

The actors who initiated the changes with tectonic affects within the Phase A of construction of the Monarga House are listed in Table 4.6. The contractor, builder, foremen, and workers made 11 changes with tectonic affects, while the RONAC and the main controller made only three changes.

Table 4.5 The changes during the Phase A of the Monarga House.

Phase A	Eliminated changes	Singular changes	Multiple changes	Total number of tectonic changes	All changes	Innovative attitudes
Foundations (A2)	None	None	1 positive, 2 negatives	1 positive + 2 negative: 3	5	1 unrealized possibility causing tectonic affects.
Frame system (A3)	1 positive	None	None	1 positive: 1	3	1 unrealized possibility causing tectonic affects later
Walls, openings and, insulation (A4)	None	1 positive	1 positive, 1 negative	2 positive + 1 negative: 3	5	1 realized, 1 unrealized possibility, all causing tectonic affects
Roof, heat insulation, and pergolas (A5)	1 positive	1 positive, 3 negative	2 positive	4 positive + 3 negative: 7	23	3 realized (2 with tectonic affects), 3 semi-realized, 1 unrealized with tectonic affects
Total	2 positive	2 positive, 3 negative	4 positive, 3 negative	8 positive + 6 negative: 14	36	4 realized (1 with tectonic affects) 4 unrealized (all with tectonic affects), 3 semi-realized

Table 4.6 The actors initiating changes with tectonic affects in Phase A.

Phase A	Contractor/builder/ foremen/workers	The RONAC/ the main controller
Foundations (2)	1 positive, 2 negative	None
Frame system (3)	None	1 positive
Walls/heat isolation/openings (4)	1 positive, 1 negative	1 positive
Roof/pergolas (5)	3 positive, 3 negative	1 positive
Total	5 positive + 6 negative: 11	3 positive

Note that some decisions were made collectively by all parties.

This chapter includes reflections from various theories, such as ethics, general philosophy, aesthetics, tectonics, architecture, and construction.

> *The reflections of theory* related to the changes, tectonic affects, and innovative attitudes during Phase A are:
>
> - **Ethics theories** (such as virtue ethics – competence – of Aristotle, deontological ethics and responsibility of Immanuel Kant, agreement ethics of John Rawls)
> - **Various philosophies** about power (ignorance), epistemology (evolutionary and revolutionary innovations, being open for change), metaphysics (instrumental approach, enframing, standing reserve) and poetics (meaning of attics and basements)
> - **Theories of aesthetics** (such as errors to articulate, fetishist characteristics of aesthetic detailing)
> - **Theories of tectonics** (such as hidden elements and components, innovations, logos of techne, techne of logos, heaviness and lightness, hapticity, tectonics of details, continuity with the natural environment)
> - **Theories of architecture** (such as errors, benign errors, hapticity, topophilia, monotony, meaninglessness, ambiguity, hierarchy of values, form, image, camouflage, sensational image making)
> - **Theories of construction** (standard repetitive details, innovations, management of change)
> - There are also some in-between concepts such as management of ambiguity in design.
>
> It should also be highlighted that the concept of innovation takes place within various philosophies and theories of tectonics as well as theories of construction.

5

Tectonic Affects and Innovative Attitudes due to Changes Relating to Plastering, the Electrical, and Mechanical Systems, the Ceramics, and Water Isolation

Unlike those in Chapter 4, the elements of construction in Chapters 5–7 do not have a simple order. Each one extends throughout a long time period. For example, plastering starts when the walls are built. But there are three different layers, and the application of the leaf plaster forms the last layer. Each layer is applied depending on other applications. After the application of the first layer of plaster, the marking for electricity starts. After installing some cables, the second layer of plastering is done. The third layer covers the cables completely. Plastering also has a relationship with the positioning of the window and door frames. Many activities form a chain relation with each other. The RONAC (Researcher, Owner, Neighbor, Architect, Controller) expresses this situation:

> I asked the builder to install the missing ceramic piece in the bathroom and to solve the problem in the attic before the anything else, because the fireplace can only be painted after this, and the cleaning firm and the white goods can come later. I can move into the house then.

Reflection(s) from Theory

Modern culture accepts **time as linear** and prefers to read developments through time. However, time in nature and traditions is **cyclical**. Just like the seasons they repeat themselves. If cycles overlap and are contained within other cycles in a complex way, the concept of **fractal** is used to express this situation about time.[1] **CPM (Critical Path Method) and PERT (Program Evaluation Review Technique)** graphs which are used for construction management,[2] have the complexity of fractal time and they also combine the linearity of modern time.

---

1 Max Tegmark "On the Dimensionality of Spacetime" *Classical and Quantum Gravity* 14, 1997: L69–L75.
2 Bal Chand Punmia and Krishana Kumar Khandelwal, *Project Planning and Control with PERT and CPM*. 4th Edition. (Laxmi Publications, 2017).

Tectonics as a Process in Architecture, First Edition. Yonca Hurol.
© 2025 John Wiley & Sons, Inc. Published 2025 by John Wiley & Sons, Inc.

Plastering, Painting – Changes, Tectonic Affects, and Innovative Attitudes

There is a very particular preference for the plaster of this building. This is the application of traditional leaf plaster in Cyprus. Not all plasterers can apply leaf plaster. It requires "a good hand." Since it is thicker than other types of plaster it is also heavier. It must be painted on with a brush, and it requires more paint than other plastered surfaces. This type of plaster also determines the color of the building because white shows up in this type of plaster better than any other color. Figures 5.1a–c show the first layers of plaster and the application of mesh between the reinforced concrete elements and the brick surfaces to avoid the plaster cracking due to material differences. Figures 5.1d and e show the third layer of plaster and the ceiling plaster. Figures 5.1f and g demonstrate the application of leaf plaster by an experienced plasterer and painting after the application of the third layer of plaster.

Reflection(s) from Theory

The ethics of **experience** in crafts is explained clearly by Martin Heidegger through the accumulation of changes in the product through the bodily experience in the long term. This accumulation requires a very long time spent on the production of crafts.[3] Because of this, it is not easy for beginners to have the same level of quality as experienced craftsmen. However, crafts have lost their importance because of the rise of the professions and the arts which are not based on experience. Today the legal building processes require the signature and approvement of professionals and this puts builders and craftsman into a secondary position within building sector.

Experience is one of the foundations of **authority** in cultures. Remember the term 'master' in the master and apprentice system of education which took a long time. Experienced people who have lived for a long time, had authority over younger people.[4] Modernity is against this type of authority and scientific knowledge has become more important than experience. Critical approaches also relate authority to authoritarian personality which is demand of authority without any basis.[5]

---

3 Martin Heidegger, *Being and Time*, Trans. John Macquarrie and Edward Robinson. 25th Edition. (Blackwell Publishing, 2005[1927]).
4 Claire Blencowe et al., "Authority and Experience" *Journal of Political Power* 6, no. 1, 2013: 1–7. DOI: 10.1080/2158379X.2013.774973.
5 Theodor Adorno et al., *The Authoritarian Personality*. (Verso Books, 2019[1950]).

Plastering, Painting – Changes, Tectonic Affects, and Innovative Attitudes | 151

Figure 5.1 (a) First layer of interior plaster. (b) First layer of exterior plaster. (c) Application of mesh to the façade where there is material change. (d) Third layer of exterior plaster. (e) Ceiling plaster. (f) Leaf plaster on façades. (g) Painting.

> **(Continued)**
>
> Since crafts were downgraded while the Monarga House was being built, there was no authority involved in the application of leaf plaster which requires experience. It was applied very modestly.

Today I visited the site to see the application of leaf plaster. Almost half of it has been done. The builder was worried about a part of the façade which has to have leaf plaster, but has been treated as if it is not going to have leaf plaster.

Changes During Plastering and Painting

- The project specified leaf plaster on most of the outside walls of the building. However, there were doubts about using this plaster together with heat insulation of the structural elements. As the heat insulation was canceled, leaf plaster could be used. However, this type of plaster was too costly. The façades of the house were redrawn to show where leaf plaster and normal plaster should be applied. Terraces were plastered in the normal way and leaf plaster was retained in other areas. This eliminated rough surfaces on the terraces and added some variety to the façades. This is one of the issues that changed several times during construction.

Tectonic Affects due to the Changes During the Application of Plastering/Painting

Plastering was done by an experienced builder who managed the activity well and made a considerable contribution to the construction. Only one tectonic affect was caused during this activity and that was initiated by all actors of the building team (Table 5.1).

When we were there (after the painting finished), the pigeon chick was still sitting on the AC (air conditioning) unit and was very stressed. Workers were careful with it.

Innovative Attitude During Plastering and Painting

- The problem which could have triggered an innovative attitude during plastering, is the problem of heat insulation which appeared during the construction of the reinforced concrete frame system and continued during the construction of the brick walls. This problem did not lead to any innovative attitudes, but it resulted in positive tectonic affects because of the application of the leaf plaster.

Electrical System – Changes, Tectonic Affects, and Innovative Attitudes | **153**

Table 5.1 Tectonic affects due to changes during the application of plastering and painting.

The change in 6	The symbol	The tectonic affect(s)	Responsible person/ contributor	Related changes	Innovative attitude
Application of leaf plaster on certain walls.		Hapticity, continuity with the vernacular/ traditional architecture.	All members were involved.	Many	Unrealized possibility

Eliminated changes and problems are shaded in light gray in the table. Changes which caused another change or originated from another change are shaded in darker gray. Changes that did not cause any other changes are unshaded. Symbols reflect not only the type of change, but also the associated feelings. Lighter colors in symbols reflect positive feelings.

Electrical System – Changes, Tectonic Affects, and Innovative Attitudes

The installation of electrical systems takes almost as long as the whole construction process after the frame system and walls are in place. Figures 5.2a and b demonstrates the placement of the pipes through which the electric cables will pass through the reinforced concrete slabs. Figures 5.2c and d show where the electric cable pipes will be installed after the application of the first layer of plaster. Figures 5.2e and f show the electricity connection for the AC and the electricity pipes which run under the floor. Figures 5.3a–c show further development in the installation of the electrical system. Figure 5.3d shows the trench through which the house is connected to mains. Figure 5.3f shows the electricity manhole. Figures 5.3e and g show the building with the lights on and the electricity support system. Figure 5.3h shows the placement of the ACs. Finally, Figures 5.3i and j show the two lamp posts at the entrance of the house and the internet connection. As seen from these figures, the electrical system is installed after the application of the first layer of plaster and continues almost to the end of the construction.

Reflection(s) from Theory

Eduard Sekler's *atectonic* concept is about the **hidden construction components and elements** in buildings.[6] Similarly, Kenneth Frampton's concepts of *tectonic form* and *ontological construction* cover construction components and

---

(Continued)

---

6 Eduard Sekler, "Structure, Construction, Tectonics." In *Structure in Art and Science*, Ed: G. Kepes. (Wordpress, 1965): 89–95, accessed August 18, 2018. https://610f13.files.wordpress .com/2013/10/sekler_structure-construction-tectonics.pdf.

154 *5 Tectonic Affects and Innovative Attitudes*

Figure 5.2 (a) Electricity pipes placed in the ceiling formwork. (b) Electricity pipes passing through the concrete ceiling. (c) Indication for placement of electricity cable pipes inside the building – after applying the second layer of plaster. (d) After applying the first layer of plaster. (e) Electricity connection for ACs. (f) Electricity pipes on the floor.

Electrical System – Changes, Tectonic Affects, and Innovative Attitudes | 155

Figure 5.3 (a–c) Development of the electricity system. (d) Digging for the electricity connection from the house. (e) The house with electricity. (f) The electricity manhole. (g) The electricity support system in the attic. (h) Position of ACs. (i) Lamp posts in the garden. (j) Connection to the Internet.

> *(Continued)*
>
> elements which enable **reading the logic of their construction.**[7] However, these theories do not consider the construction logic of the electrical and mechanical systems. Many parts of electrical and mechanical systems are hidden, and it is usually not possible to read the logic of their construction when it is completed.

Immediately after getting electricity, I can call the plumber to install the bathroom. The carpenter also needs electricity to generate heat to remove the varnish on the timber posts. Electricity is becoming very urgent and critical, and nobody knows how to deal with it properly. Stressful.

Changes During the Installation of the Electrical System[8]

- Although the plugs, lamps, and so on, were placed in accordance with the RONAC's wishes, they were awkward to use because the electrician placed them too far from the edges (#13).
- The AC placements were not shown in the preliminary project. Therefore, the electrical engineer placed them without consulting the RONAC. This caused some functional and also aesthetic problems later. One of the ACs is in the worst place – the south terrace.
- The tops of two of the ACs were too close to the eaves and pigeons and sparrows started using them for nesting. We had to put wire mesh behind the ACs to avoid this. However, this did not work. Currently this problem has been solved by a carpenter applying framed nets to these ACs in a distinct way.
- The position of the electricity manhole was not considered during the preliminary design phase. Finally, it had to be placed adjacent to the entrance terrace. This caused some aesthetic problems. "This pathway will also make the electricity manhole look better. I remember that I was worried about the placement of this manhole. Because it is very close to the entrance."
- An electricity support system was added to the house. However, the placement of its solar panels became a problem. There was not a sufficiently large area on the roof.
- The main controller insisted on adding a cooker hood to the kitchen and this was accepted by the RONAC.

---

7 Brian Kenneth Frampton, "Rappel a l'Ordre: The Case for the Tectonic." In *Labour, Work and Architecture*, Ed: K. Frampton. (Phaidon, 2002): 91–103. Brian Kenneth Frampton, *Studies in Tectonic Culture – The Poetics of Construction in Nineteenth and Twentieth Century Architecture.* (MIT Press, 1995).

8 Hashtags (#) indicate mistakes which could have been identified and solved by experienced foremen or technicians.

Electrical System – Changes, Tectonic Affects, and Innovative Attitudes | **157**

- Although the original project catered for electricity in the garage, electricity was not available there. Instead, the RONAC put a solar lamp inside the garage.
- It is difficult to have a camera installed in the front garden because there is no electricity there. However, chargeable portable cameras could be used which do not need electricity.
- The method of illuminating the garden was not defined in the project. This was done later by placing many solar lamps throughout the garden.

Reflection(s) from Theory

The history of electricity goes back to the nineteenth century and the introduction of mechanization in buildings goes back to the beginning of the twentieth century. The nineteenth and twentieth centuries also saw the emergence of the **professional approach** to building production as well as new professions such as civil, mechanical, and electrical engineering. However, despite these radical changes, architecture is one of the oldest human activities.[9] Other **building systems** are designed with the **coordination** of architectural practice. Consideration of the relationship between these systems, the building, and the architectural character of the building is essential during this process, and this can be achieved through **collaboration** between different professions during the building activity.[10] Communication, coordination, and collaboration during the construction of the Monarga House were not effective.

The RONAC realized that the project drawings used by the electrician were not part of the application project. The electrician informed the RONAC that the drawings he had contained fewer ACs, and so on. These sheets were not the same as the RONAC's. Since the RONAC shared the electronic versions of the application project with the contractor, somebody made some changes to the electrical engineering project.

Reflection(s) from Theory

According to Albert Chan and Emmanuel Owusu, forms of **corruption** in the construction industry include **fraud (falsification)**.[11] Making changes to the agreed project of the Monarga House can be regarded as fraud.

---

9 George Barnett Johnston, *Assembling the Architect: The History and Theory of the Professional Practice*. (Bloomsbury Visual Arts, 2020).

10 Ute Poerschke et al., "BIM Collaboration Across Six Disciplines." In *Proceedings of the International Conference on Computing in Civil and Building Engineering*, Ed: W. Tizami. (Nottingham University Press, 2009): 575–671.

11 Albert P.C. Chan and Emmanuel K. Owusu "Corruption Forms in the Construction Industry Literature Review" *Journal of Construction and Engineering Management* 143, no. 8, 2017: 04017057.

Tectonic Affects due to Changes During the Placement of the Electric System

The electrician was not particularly communicative. He was trying to complete his work according to his version of the project. There were four changes causing tectonic affects during this phase of construction. Two of these changes were initiated during the application project and by the contractor (Table 5.2).

> When the lamp posts at the entrance were installed it was lovely. I turned them on, and they made me feel happy. Then I played music and enjoyed it.

Innovative Attitudes During the Application of the Electrical System

- The placement of ACs too close to the eaves of the roof attracted birds. A typical framing with nets could not be applied to these ACs because the eaves are

Table 5.2 Tectonic affects due to changes during the installation of the electrical system.

The change in 7	The symbol	The tectonic affect(s)	Responsible person/ contributor	Related changes	Innovative attitude
Bad placement of ACs.		The outdoor part of the largest AC was put on the terrace with a sea view. It also obstructs the movement of one of the shutters. Bad detail.	Application project and the RONAC	0	None
Bad placement of the electricity manhole.		The electricity manhole was placed beside the entrance terrace. Bad detail.	Application project/ contractor	0	None
Addition of the cooker hood to the kitchen.		Silver color cooker hood gave a more modern tectonic outlook to the kitchen.	Main controller	0	None
Addition of solar lamps in the garden.		Having lights turning on automatically at nights gave a live character to the house.	The RONAC	0	None

Eliminated changes/problems are shaded in light gray in the table. Changes which cause another change or originate from another change are shaded in darker gray. Changes that did not cause any other changes are unshaded. Symbols reflect not only the type of change, but also the associated feelings. Lighter colors in symbols reflect positive feelings.

slanting. The carpenter had an innovative attitude and covered the top parts of the frames with timber surfaces. These timber surfaces are in harmony with the timber of the roof, and they contribute to the tectonic affects caused by timber elements and components in the building. However, this was done a few years after the construction was completed.

Reflection(s) from Theory

The concept of *tectonics* originates from the Ancient Greek noun *tecton* (τέκτων) which means artisan, craftsman, and especially carpenter or wood-worker. Woodwork is emphasized because of the **playful character of timber**. Because of this it is associated with the human mind.[12]

Mechanical Systems – Changes, Tectonic Affects, and Innovative Attitudes

The RONAC gave specific importance to the mechanical systems in order to have no problems in the future. She chose a 4-ton water tank at the beginning of the application project. She asked for a quality pump and a booster. She also selected bathroom and kitchen fixtures such as taps and showers very carefully. However, although she and the main controller followed the process quite closely, there were many problems at the end.

Reflection(s) from Theory

Electrical and mechanical systems in older buildings are usually poor. This is because of the use of **old systems** and **improvements in standards**. Mechanical and electrical engineers as well as the firms which sell these systems follow the changes in the systems and standards.[13] The RONAC preferred to have up-to-date electrical and mechanical systems in the Monarga House.

The placement of mechanical systems lasted for most of the construction period. It started with the installation of the electricity system and ended not long before the construction ended. Figures 5.4a–c show the black water pipes with the electricity pipes on the floor and some outlets in the kitchen and bathroom. Figures 5.4d and e show the water and electricity pipes on the floor and that

---

12 Gaston Bachelard, *The Poetics of Space*, Trans. M. Jolas. (Beacon Press, 1994[1958]).
13 Frederick S. Merritt and Jonathan T. Ricketts, *Building Design and Construction Handbook*. 6th Edition. (McGraw-Hill Professional, 2000).

160 | *5 Tectonic Affects and Innovative Attitudes*

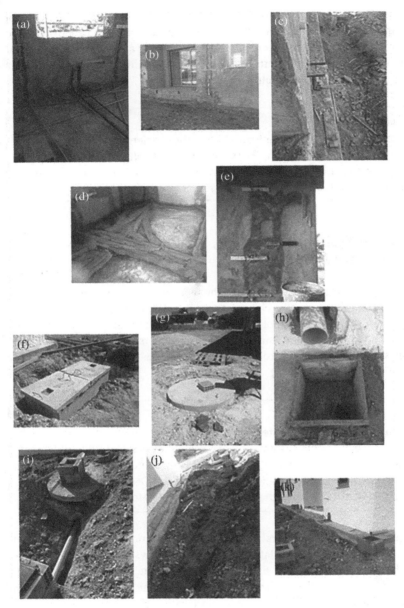

Figure 5.4 (a) Water pipes placed among the electricity pipes. (b) Installation of outlets for dirty water and sewage from the kitchen and bathroom. (c) Some of the outlets. (d) The water and electricity pipes covered with concrete. (e) Electricity pipes on the walls covered with plaster. (f) Installation of one of the two septic settling tanks. (g) The top cover of the main septic tank. (h) The toilet manhole. (i) Connection of the pipe to the main septic tank. (j) Pipes covered with concrete. (k) Manholes adjacent to the building and manholes in the garden.

the walls are covered. Figures 5.4f–h show one of the septic settlement tanks, the top cover of the main septic tank and the toilet manhole. Figures 5.4i–k show the pipe connection to one of the septic tanks, concrete applied to the septic pipes, and most of the manholes in the building. Figures 5.5a–c present various stages of the construction of the bathroom. Figures 5.5d–f demonstrate the placement of outdoor water pipes, concrete applied to these outdoor pipes, and the placement of the water tank. Figures 5.5g–i demonstrate the drain filter for the shower cabin, the water pump and booster pump, and the completed bathroom.

Changes Made During the Placement of Mechanical System Elements[8]

The mechanical engineer says that the contractor of my house will do all sorts of things to buy cheaper products. That worries me because the contractor says we have to buy the ACs now. I must be very careful because he might send us one document, but the product might be different. He did not even know the BTU (British Thermal Unit) of the ACs today. How he is going to buy them without looking at the project details, I do not know.

Reflection(s) from Theory

According to Ian McKinnon, **unnecessary, inappropriate, and poor-quality goods or services** are indications of **corruption**.[14] Since every item is defined in the project specification for the Monarga House, trying to buy cheaper products despite their low quality, is an example of corruption.

- The locations of septic tanks were not defined in the application project. However, their placement is good because they are not at the front of the west terrace. This is an aesthetic contribution of the contractor to the building.

The truck drivers did not have anything to cover these holes (the well and sewage holes) and I was worried. A child or an animal could fall into them. So, they found pieces of wood and some branches of trees to show where the wells are. This morning, they came again, probably to cover the holes, but I haven't checked yet.

---

14 Ian McKinnon, *Corruption in Construction: How to Tackle this Industry-wide Problem*, 2020, accessed February 16, 2025. https://www.chas.co.uk/blog/tackling-corruption-in-construction/.

162 | *5 Tectonic Affects and Innovative Attitudes*

Figure 5.5 (a) Installation of some of shower and toilet fixtures. (b) The shower and toilet fixtures are plastered and leveling concrete applied to the sloping bathroom floor. (c) Leveling concrete applied to the sloping floor of the shower cabin. (d) Garden pipes installed. (e) Garden pipes covered with concrete. (f) Water tank placed on the concrete base. (g) The shower drain filter. (h) The water pump and booster pump are installed. (i) The completed bathroom.

Mechanical Systems – Changes, Tectonic Affects, and Innovative Attitudes | **163**

Reflection(s) from Theory

There have to be various **safety measures on construction sites**. The major precautions concern:

- working at height
- site traffic
- moving goods safely
- **dangers due to groundwork**
- demolitions/alterations
- occupational health risks
- dangers due to electricity
- slips and trips
- working in confined spaces
- prevention of drowning
- requirement to use protective equipment
- work affecting the public
- requirement of monitoring and reviewing.

Groundwork precautions should prevent **people and vehicles from falling into .excavation hole.**[15]

Health and safety precautions were not followed fully during the construction of the Monarga House.

I found out that they brought the pipes for the septic tanks. I am worried about their dimensions. The main controller said they should be 4-inch and 3-inch pipes. Then it should be alright. However, the number of manholes was problematic, and the main controller asked to have manholes for all connections. This was done.

Reflection(s) from Theory

Contractors' attempts to increase their interest may include decreasing the number of manholes although it is functionally better to have all of them. Martin Heidegger and Jacques Ellul explained the **instrumental approach** to the world as an orientation towards the scientific approach to achieve economic

---

(Continued)

---

15 Health and Safety Executive, *Health and Safety in Construction*. 3rd edition, Information Policy Team. (The National Archives, Kew, London, TW9 4DU, 2006).

> **(Continued)**
>
> benefit.[16] However, the discussion about the manholes between the contractor and the main controller during the construction of the Monarga House demonstrates that the contractor did not consider the scientific approach. Therefore, this idea cannot be called an instrumental approach. It is appropriate to call it corruption. According to CHAS, this definition can be regarded as **unnecessary, inappropriate, and using poor-quality of goods or services**, as well as a **reduced commitment to quality, ethics, and compliance**.

- The toilet manhole was built higher than expected because the hole for the passage of the toilet pipe from the foundation tie-beam level had not been made (#14). The necessary preparations were not done for a lower manhole. This created an ugly appearance at the west side of the house.
- The other manholes in the garden were also too high. "I saw in the morning that the manholes are almost complete. They worried me a bit because many of them were higher than expected." This was because the builder and technician thought they would need to be filled with earth (#15). However, the RONAC wanted to see the rocky surface of the site as much as possible. The RONAC had to have the manholes lowered later. Still, rocky ground is only visible in a very few places in the garden after completion of the building.

> *Reflection(s) from Theory*
>
> **The contractors and builders looks as if they are trying to increase their interest** in any building activity. Considering the construction of the Monarga House, it can be said that digging less deep, decreasing the height of floors, having a lower roof, and having fewer manholes are examples of such behavior. Since such behavior runs counter to the architectural project, it can be categorized as systematic violation of the agreement with the client. Therefore, it can be seen as **unjust, due to violation of the agreement ethics**.[17]

---

16 Martin Heidegger, "The Question Concerning Technology." In *The Question Concerning Technology and Other Essays*, Trans. William Lovitt. (Garland Publishing, 1977[1954]a): 3–35. Martin Heidegger, "The Age of World Picture." In *The Question Concerning Technology and Other Essays*, Trans. William Lovitt. (Garland Publishing, 1977[1954]b): 115–154. Martin Heidegger, "Science and Reflection." In *The Question Concerning Technology and Other Essays*, Trans. William Lovitt. (Garland Publishing, 1977[1954]c): 155–182. Jacques Ellul, *The Technological Society*, Trans. J. Wilkinson A. (Alfred Knopf and Random House, 1964). Jacques Ellul, *What I Believe*, Trans. G. W Bromiley. (Eerdmans, 1989).
17 John Rawls, *A Theory of Justice*. (Harvard University Press, 2009[1971]).

- The RONAC found out much later that having brick manholes can cause problems in the long term because they are under the ground and liable to water damage.
- The sewage vent pipe was not installed. The RONAC had to deal with this later. One of the experienced technicians noticed the problem and inserted the pipe (#16).
- The manhole for the water meter was too small (#17). This damaged the water meter and its manhole had to be enlarged later.
- The water tank was changed twice: from 4 tons to 2 tons during the application project and from 2 tons to 5 tons during construction. "The contractor accepted a larger tank. He was worried about the shape and size of it. In the end an awkwardly shaped tank was delivered. It is circular and vertical. It will not fit into the area in front of the manholes. The radius of the tank is around 2 m but the place around the manholes is 1.7 m." This change caused an aesthetic problem on the west façade. Its position was also different from that of the original project because of the tank's radius and this was creating a problem with the manholes. The tank was in front of the kitchen in the original project. However, it had to be moved in front of the bathroom. Since that area could have drainage problems, the concrete base of the tank was placed in such a way that water can flow from both sides. Also, the position of the tank was reorganized by the builder and the plumber. Rather than having many pipes around it, the builder and the plumber had the clever idea of using the existing underground pipework for the same purpose. This changed the direction of the head of the tank. This was better because too many things were getting too close to each other: the bathroom window, the lighting element, the AC piece, and the head of the tank. Now the head is directed away from the building, and it does not block anything. This was the contribution of the plumber. Later the RONAC had the tank covered with a timber screen. The issue about the water tank caused several changes during the building process, because of the problem created during the application project by changing the water tank from 4 tons to 2 tons.

Reflection(s) from Theory

Why do architects **not include the builders and technicians in the design process**? Why not have a more **inclusive collaboration**? There are two different answers to these questions.

One of them is about **professionalization** and this applies to architects as well as engineers. From the classical period onwards, mental activities have been regarded as separate and more important than bodily activities. Between 400 BC and 250 BC there was a significant **degradation of materiality**

(Continued)

166 | *5 Tectonic Affects and Innovative Attitudes*

> *(Continued)*
>
> to **support the mental activities** of humans such as philosophy.[18] This division has become more evident after the end of the nineteenth century together with the emergence of related professions based on scientific and technological knowledge. The emergence of the concept of aesthetics during the seventeenth century also had a more abstract character in comparison to ancient art which was more like craft. Architecture became a profession which also has connections with twentieth-century art.
>
> A second answer can be given to the above questions by considering the relationship of architecture to art. Sociologist Hasan Ünal Nalbantoğlu wrote that architects self-conceptualize as artists even when they work as draftsmen. He thought that **being an artist is the spontaneous ideology of architects.**[19] According to Peggy Deamer, seeing themselves belonging to the "creative class," **architects do not conceptualize themselves as "workers"** and the majority do not even seek their rights.[20]
>
> Being a professional and an artist and not a worker, may stop architects from accepting construction workers' and builders' attitudes as creative and/or innovative.

- The cover of the water tank was blue, and this was not part of the color scheme of the house. The RONAC had to put a whitish nylon-based cloth over the cover to hide this problem.
- The contractor did not want to buy the booster pump although it was included in the project. This was not accepted by the RONAC, and the booster pump was fitted.
- The boiler in the attic was not the same quality as suggested in the mechanical engineering project. The RONAC was unable to change this later.
- The solar panels on the roof were not the same quality as suggested in the mechanical engineering project. The mechanical engineer has said that the metal surfaces of the sun collectors are not aluminum which is not good.

---

18 Stanford Encyclopedia of Philosophy, *Episteme and Techne*, 2021, accessed February 21, 2025. https://plato.stanford.edu/entries/episteme-techne/.

19 Nalbantoğlu refers to Louis Althusser for the concept of spontaneous ideology. Althusser used this concept for scientists. Louis Althusser, *Philosophy and the Spontaneous Philosophy of Scientists*. 2nd Edition. (Verso, 2011[1974]). Hasan Ünal Nalbantoğlu, "Yaratıcı Deha: Bir Modern Sanat Tabusunun Anatomisi (Creative Genius: The Anatomy of a Modern Art Taboo)." In *Çizgi Ötesinden: Üniversite: Sanat: Mimarlık*, Ed: H. Ü. Nalbantoğlu. (ODTÜ Mimarlık Fakültesi Yayınları, 2000): 85–104.

20 P. Deamer (ed.), *The Architect as Worker – Immaterial Labor, the Creative Class and the Politics of Design*. (Routledge, 2015).

Also, the thickness of the glass surface was not acceptable. These will go wrong very soon. Also, the boiler was not welded properly and will soon cause problems. The RONAC was unable to change these. Since the placement of the solar panels was not thought through during the preliminary and application projects, their positioning was not satisfactory.

Reflection(s) from Theory

Standards raise the level of professional practice. They provide an alternative for poor production through an increase in efficiency and a reduction in costs. They also create confidence in the industry because of the reduction of waste and better health and safety. There can be standards of building materials and systems (such as ISO standards) as well as building codes which specify the building quality requirements for different countries. **Building codes** can be prescriptive or performance based. If the codes are prescriptive, they force architects and engineers to design in certain defined ways. However, if they are performance based, architects and engineers can design freely and demonstrate the performance of their design. Both standards and building codes evolve over time and they also follow the values of their time. For example, sustainability provides important criteria for contemporary standards and building codes.

As an example, Steve Thomas' book *Building Code Essentials* is based on the 2015 International Building Code of the USA.[21] Not following standards and building codes can be regarded as **out of date** within contemporary building design and production.

- There is a 1-ton tank and a boiler in the attic. Since these create a risk of flooding, an isolated pool was built under them in the attic.

  "I called the firm and talked with the architect and asked him to send me the remaining items (for the bathroom) that I selected from their shop. They will be here tomorrow morning. I am very excited." The bath drain filter was too long, and it increased the width of shower cabin towards the window, so it was changed. The firm's architect identified this problem (#18). However, the replacement had a broken cover, and it was filled with earth and mortar. I asked for the drain filter to be changed again. This problem was solved by a plumber in the end.

---

21 Steve Thomas, *Building Code Essentials*. (ICC International Code Council, 2015).

> *Reflection(s) from Theory*
>
> The philosopher Martin Heidegger uses the concept of **care** as an important feature of Dasein (the human being who is *there* – who is bounded to a place, to the world). Dasein is **thankful for the contributions** of others for his/her existence. Dasein **does not use others as tools** for his/her own benefit. Dasein knows that we are all **thrown into this world** and our presence is temporary. The life of humans may end but the **world persists**. Dasein acknowledges that **every human being has possibilities**, and s/he may also fall by the wayside, mindlessly following the calculative attitudes of other people. Care is about **authenticity** which requires **protecting the world and all beings**, protecting the possibilities of other humans, and protecting one's own self despite knowing that life is temporary.[22]
>
> The above event relating to the drain filter in the Monarga House reflects two opposing behaviors in relation to care: the care of the firm's architect for the design of the bathroom and the careless behavior of one of the workers who broke the cover of the drain filter and let it be filled with earth and mortar.

- The montage detail of the closet using plaster was an issue for discussion. The problem was solved by the contractor.
- The garden taps were connected to the city water supply at the beginning. But these taps were then connected to the tank to avoid a water shortage in the garden if the water supply was cut.
- One of the shutters cannot be fully opened and is attached to its stopper because of one of the AC units. The technician said he might come later to change the placement of this unit. It needs approximately 20 cm of movement to the right. However, this never happened (#19).
- The RONAC decided to have a water filtering system in the house to obtain drinking water directly from one of the taps. This will need to be serviced twice a year. Since there is no place for water bottles in the kitchen this system has become very useful and practical.

Most of the items within the mechanical engineering project were either changed or tried to be changed by the contractor, although he had seen the mechanical engineering project before signing the contract with the RONAC.

---

22 Heidegger, *Being and Time*.

Tectonic Affects due to Changes During the Placement of the Mechanical System

There were serious problems in the application of mechanical systems. The technician who installed most of the mechanical systems was in the contractor's team. There were six changes, which caused tectonic affects, within this phase (Table 5.3).

> While the plumber was smoking, his son went inside but ran out again in a panic. We all ran in to see that both the kitchen and bathroom were flooded. I started worrying about the timber door frames and the cupboards in the kitchen. I rushed home to get some cloths and buckets. The plumber should have put some buckets under these taps when he was working on them. He didn't ... I got angry with him and told him that he did not know what he was doing. I was sad about this later because his son was there. We all mopped up the water. My dog was rolling in the water and making our job more difficult.

> The water from the taps was a bit yellow. One of the builders said they had forgotten to wash the tank before filling it.

Innovative Attitudes During the Application of Mechanical Systems

There were four possibilities for innovative attitudes.

- The problem that appeared during the installation of the mechanical systems is the continuation of the one that first occurred during the construction of the foundations. This relates to the placement of a high toilet manhole near the west façade because a hole was not drilled in the foundation tie-beams. This problem did not trigger any innovative attitudes, and it causes negative tectonic affects.
- There was the possibility of innovative attitudes during the montage of the closet to the wall and this happened.
- There were some innovative attitudes from the builder and plumber during the installation of the water tank.
- Constructing an isolated pool beneath the 1-ton tank in the attic as a precautionary measure against flooding was an innovative attitude.

Table 5.3 Tectonic affects due to changes during the installation of mechanical systems.

The change in 8	The symbol	The tectonic affect(s)	Responsible person/ contributor	Related changes	Innovative attitudes
Determination of the placement of septic tanks.		This prevented the manholes from blocking the west terrace.	Contractor, the RONAC	0	None
All the manholes and especially the toilet manhole were higher than expected.		The toilet manhole caused a disturbance to the back façade of the building because of bad detailing.	Problem by: Contractor had to be accepted by the RONAC	Many	Unrealized possibility
Having a 2 ton water tank.		Although a smaller tank would be tectonically advantageous, it contradicts the project's requirements and its placement also posed issues.	Problem: Application project Solution: the RONAC	Two	None
Decisions about the placement and capacity of the main water tank: From 2 tons (negative) to 5 tons (positive).		One 5-ton, circular and vertical water tank with a blue cover was placed in front of the bathroom. This tank was later covered with timber shields which gave it a haptic character.	Solution: the RONAC, technician	Many	None
The vent pipe of the toilet was added later.		Since the vent pipe was added later, it became a visible pipe near the back façade.	Problem: Contractor Solution: A different technician	0	None
The manhole for the water meter had to be enlarged.		This affected the garden negatively because this manhole is visible everywhere in the garden.	Problem: Builder Solution: Another builder	Two consecutive actions	None

Eliminated changes/problems are shaded in light gray in the table. Changes which cause another change or originate from another change are shaded in darker gray. Changes that did not cause any other changes are unshaded. Symbols reflect not only the type of change, but also the associated feelings. Lighter colors in symbols reflect positive feelings.

Ceramics – Changes, Tectonic Affects, and Innovative Attitudes

Ceramic tiles also had an important place in the design of the Monarga House. A colorful ceramic type was selected, and it was clear that the floor covering would be dominant. These ceramics were intended to cover the whole floor (including the terraces) and be placed on some walls in the kitchen and bathroom. The type of ceramic was selected because it could be used inside, outside, and in wet spaces.

Reflection(s) from Theory

The treatment of surfaces in and on buildings has a specific importance in the tectonics of historic Islamic buildings. The walls of buildings – the material parts – are seen as continuations of earth while the spaces in and around the buildings are seen as continuations of the sky. This indicates that the surfaces in and on buildings are the boundaries between earth and sky. Therefore, an **articulation of the surfaces which separate the earth and sky**, by carving them or by covering them with calligraphy, represents care for the relationship between earth and sky.[23] At this point it is also necessary to remember Martin Heidegger's concepts of earth, sky, mortals, and divinities. Earth represents the material world and the sky represents the human mind with all its capabilities. Mortals are the temporary humans and divinities are the ancestors and culture of humans.[24]

Although the RONAC had no intention of giving her house an Islamic character, she was aware of the treatment of surfaces in Islamic buildings. The importance of the texture of ceramics on the floor and walls as well as the use of leaf plaster which adds texture to the outside walls, can also be seen as subconscious decisions about the tectonic affects of the Monarga House.

> I think the ceramics gave me huge freedom regarding color. Every color is there.

The installation of ceramics also took a long time lasting from the second layer of plastering up to almost the end of construction. Figures 5.6a and b show the application of the leveling concrete to the interior part of the building and the placement of ceramics in the interior spaces. Figures 5.6c–e show ceramic tiles on

---

23 Yonca Hurol and İbrahim Numan, "Rethinking Islamic Anatolian Space." In *Mediations in Cultural Space – Structure, Sign, Body*, Ed: J. Wall. (Newcastle: Cambridge Scholars Publishing, 2008): 15–29.
24 Martin Heidegger, "Building Dwelling Thinking." In *Poetry, Language, Thought*. (Harper and Row, 2001[1951]).

5 Tectonic Affects and Innovative Attitudes

Figure 5.6 (a) Leveling concrete placed in the interior spaces. (b) Ceramics placed in the interior spaces. (c) Ceramics applied to the walls of interior spaces. (d) Ceramics applied to the terraces. (e) Ceramics applied to the relevant outdoor surfaces.

a kitchen wall and in the outdoor spaces. The indoor and outdoor ceramic tiles were at the same height in the preliminary design to achieve greater continuity between the indoor and outdoor spaces. However, this was changed during the application project to prevent water from entering the building.

> *Reflection(s) from Theory*
> **Careful consideration of context** is essential in architectural **ethics**. Different considerations of the architectural context can be seen from the ontological point of view as separations and continuities.[25] **Continuity** is poetical and the analytical mind separates things from each other. Poetic imagery relates things to each other to provide continuity.[26] Providing some continuity might

---

25 See Charles Sanders Peirce for the concept of ontology. Charles S. Peirce, "The Principles of Phenomenology." In *Philosophical Writings of Peirce*, Ed: J. Buchler. (Dower Publications, 1955): 74–97.
26 Anthony C. Antoniades, *Poetics of Architecture – Theory of Design*. (Van Nostrand Reinhold, 1992).

Ceramics – Changes, Tectonic Affects, and Innovative Attitudes | **173**

> be essential for achieving **meaning** in contemporary tectonics. **Separations** are usually achieved through **transitions** which avoid sudden changes in architecture. There can be continuity with the natural, historical, and urban contexts. The design of the Monarga House considers **continuity with the natural context**, and continuity between indoor and outdoor spaces is a part of this approach.

Many problems occurred during the installation of the ceramics because the ceramic worker was a civil engineering migrant who was not experienced in ceramics. The later phases of ceramic work had to be done by another ceramic worker.

Changes During the Placement of Ceramics[8]

- The wall ceramics were selected later, and color selection was difficult. Since wall tiles and floor tiles were to be used together on the walls, their size and thickness were also a problem. Finally, with the help of an interior designer friend the RONAC found a cream-colored ceramic. Later a shiny whitish ceramic was added. There were three types of ceramics for the bathroom and kitchen walls. The kitchen and bathroom ceramics were changed twice because of the bad placement. This was requested by the RONAC and the main controller.
- The height of ceramics in the kitchen and bathroom was another important issue because the heights of the doors, windows, and cupboards were different. Therefore, ceramics started to play the important role of providing continuity between these components and elements. Because of this, the kitchen and bathroom were redesigned and changed twice. Having three types of ceramics was another factor making these changes. "I started to think that the ceramic color (in the kitchen) could have been lighter. It looks a bit gloomy now. We shall see. I am worried about the color problems in the bathroom too. There are three types of ceramics there."
- Although the RONAC planned to have ceramics fitted to the ceiling at the beginning, she changed her mind later because the ceiling was too high, and the tiler warned her that this might not look nice (#20).

> *Reflection(s) from Theory*
> Thomas Telford (1757–1834) who is known as the pioneering civil engineer **continued to design his projects while building them**. The Craigellachie Bridge, in Scotland, is a product of such a design process. This bridge is one of the

(Continued)

> **(Continued)**
>
> first modern structures using modern structural materials. However, the structure of this bridge does not have any similarity to contemporary categories of structures. It represents the transition period from traditional structures to modern structures.[27]
>
> The Monarga House was not designed in this way; however, the design of ceramic surfaces was realized while installing them.

- The terrace edges were covered with a cream-colored marble. This was not previously planned. However, this was the contribution of the contractor to the tectonics of the house.
- The detail between the wall and floor ceramics in the kitchen and bathroom was wrong (#21). This correction was requested by the main controller. He said that the floor ceramics should continue under the wall ceramics to touch the wall. The gap between the tiles could be filled afterwards. Otherwise, there would be empty space between these tiles and water would get in.

> *Reflection(s) from Theory*
>
> **Water problems** in buildings are varied and they are a major (maybe the most important) source of **deterioration**. Water problems may occur due to mistakes in façade and roof design, and drainage and mechanical equipment problems. Water can penetrate many building materials and create channels. It can cause deterioration of materials as well as **rot and mold** which affect the health of human beings.[28] The concept of sick building syndrome covers such problems too.
>
> Some serious water problems were solved during the design and construction of the Monarga House. Solutions include the presence of a drainage well, site drainage, the isolation of the roof, two rain chains, the isolation of the building from water coming from the ground, the isolation of walls and floors of wet spaces, the ceramic details from water, the slope of outdoor surfaces, the height of the ground floor with respect to the level of terraces, the construction of a pool under the water tank and the boiler in the attic, and the position of a water exit pipe from the attic. There are no current water problems in the building.

---

27 Rowland Mainstone, *Developments in Structural Form*. (Routledge, 2001). David P. Billington, *The Tower and Bridge*. (Basic Books, 1983).

28 William Rose, *Water in Buildings – An Architect's Guide to Moisture and Mold*. (John Wiley and Sons, 2005).

Ceramics – Changes, Tectonic Affects, and Innovative Attitudes | **175**

- A foreman added an edge detail element to the outdoor ceramics which were protruding. He said this would prevent them from being broken. This is his contribution (#22).
- There is an ongoing problem with the floor ceramics inside the house. There are some holes in them. These holes occurred when some workers carelessly threw down some pieces of stone from the attic. These holes must be filled in. When the RONAC searched the Internet, she found that colored resin could be used to fill in these holes. However, the ceramic firms knew nothing about this type of maintenance. Later, the RONAC had to get this resin from abroad.

Tectonic Affects due to Changes During the Placement of Ceramics

There were many problems with the placement of ceramics. However, because of the persistent demands of the RONAC the most problems were solved. There were only three changes causing tectonic affects (Table 5.4).

The application of ceramics highlights previous mistakes in construction. The angle of the bridge-like passage to the entrance of the house was wrong. This mistake caused three rectangular areas to connect to each other with an angle of approximately five degrees. When the ceramics were placed, one of the rectangular surfaces had ceramics at an angle and this looked terrible. Also, the edge tiles were sticking out at least 20 cm without any support underneath them. To solve these problems, concrete triangular parts were added to that rectangle by injecting reinforcement, and the edge tiles and the ceramics were re-installed.

The passage (called the bridge in the diary) is straight now and the tiles on it are safe too. I was worried so much about those pieces.

Reflection(s) from Theory

Paying more attention to the **visible** parts of buildings in comparison to their **hidden** parts, can be discussed with respect to tectonic theories. This is because many tectonic theories ask for the visibility of structural and construction elements/components and understandability of the logic behind their production. If building parts are hidden, Eduard Sekler evaluates them as **atectonic**.[29] If the construction logic of building parts is not understandable, Kenneth Frampton does not evaluate it as **ontological**.[30] These theories

---

(Continued)

---

29 Sekler, "*Structure, Construction, Tectonics.*"
30 Frampton, "Rappel a l'Ordre"; Frampton, *Studies in Tectonic Culture.*

Table 5.4 Tectonic affects due to changes during the placement of ceramics.

The change in 9	The symbol	The tectonic affect(s)	Responsible person/contributor	Related changes	Innovative attitudes
The wall tiles in the kitchen and bathroom were changed twice because of the wrong placement of tiles. Rework.		Non-continuous lines and different levels for tiles create negative tectonic affects because they look awful, and they are unreasonable.	The RONAC, the main controller	Twice	None
Tiles were changed in the kitchen and bathroom to create continuity between the different height levels of doors, windows, and cupboards. Rework.		This achieves continuity between chaotic elements and creates positive tectonic affects.	The RONAC	Many	None
Terraces were edged with marble of a different color.		This brings definition to the edges of terraces and causes tectonic affects.	The contractor	0	Yes, realized

Eliminated changes/problems are shaded in light gray in the table. Changes which cause another change or originate from another change are shaded in darker gray. Changes that did not cause any other changes are unshaded. Symbols reflect not only the type of change, but also the associated feelings. Lighter colors in symbols reflect positive feelings.

reflect that the **visible parts of buildings are tectonically important**. However, there are many hidden parts, elements, and components in buildings, and this is unavoidable. For example, the electric cables, pipes for the mechanical systems, bricks, inner layers of pitched roofs, all of these hidden parts of buildings usually have functional roles only, and no aesthetic roles. Paying more attention to the visible parts of buildings may indicate that the **hidden and functional parts are less important**.

For example, during the construction of the Monarga House certain hidden parts of the building, such as the hidden layers of the roof, the mechanical equipment in the attic, and heat insulation layers, were regarded as less important.

Reflection(s) from Theory

Camouflaging of something ugly may correspond to a desire to connect to others with the help of a better image.[31] The use of ceramics to soften negative tectonic affects due to different levels of windows and doors, cupboards and wardrobes can be accepted as camouflage which creates a better image. However, a desire to camouflage something unsafe or useless affects the quality of the building negatively. For example, plastering cracked beams and columns after earthquakes is not simply a matter of image. Such dangerous columns and beams are **hidden** from others largely for calculative reasons. This concealment is different from the **atectonic hiddenness** of Eduard Sekler.[32] It is also different from the invisibility of some parts of buildings, such as water isolation layers within roof layers. Actually, these layers should not be considered as hidden. They are simply **covered**.

Innovative Attitude During the Application of Ceramics

There was one innovative attitude during the placement of ceramics.

- Covering the terrace edges with cream-colored marble was an innovative approach by the contractor.

---

31 Neil Leach, *Camouflage*. (MIT Press, 2006).
32 Sekler, *"Structure, Construction, Tectonics."*

Water Isolation – Changes, Tectonic Affects, and Innovative Attitudes

> I worried about the isolations around the tie-beams (of the foundations) because they dug there again.

The application of water isolation started during the construction of the foundations and continued during the building of the roof. This part of the book explains the later phases of water isolation. Figures 5.7a and b depict water isolation on the floor and walls of the bathroom. Figures 5.7c and d show water isolation on the pavements around the building. Figures 5.7e and f show the water isolation under the water tank and boiler inside the attic. There is a pool under them, and this pool is isolated against water. There is also a pipe which is not seen clearly in the photos, to let out flood water. The plumber said there might be flooding in 10 years' time.

Changes During the Application of Water Isolation[8]

- A pool was built under the attic tank and boiler and water isolation was applied to this pool. However, this caused a problem with the heat insulation in that part of the attic. The pipe to let water out of the attic was too narrow. It was later replaced with a wider pipe (#23). None of these applications was included in the application project. They were the contributions of an architect in a construction firm, the contractor, the main controller, and two builders and plumbers. The heat insulation material for the attic was different in the application project as well. However, it was not possible to apply heat insulation to the pool. While changing the pipe, some parts of the heat insulation material also had to be dug. These problems decreased the quality of heat insulation in the attic. Decisions about the water tank and heat insulation in the attic had to be changed several times but were partially successful. Reworking took place within these stages.

Reflection(s) from Theory

The systems approach allows building design to develop in a collaborative way by thinking about all systems in a holistic way. This happens by first considering all elements in all systems and then considering their interaction with each other. Modern modeling tools such as BIM (Building Information

Water Isolation – Changes, Tectonic Affects, and Innovative Attitudes | **179**

Figure 5.7 (a) Water isolation on the bathroom floor. (b) Water isolation on the bathroom walls. (c) The first layer of water isolation at the sides of the terraces. (d) The second layer of water isolation at the sides of the terraces. (e) An isolated water pool being prepared for the tank and boiler in the attic. (f) Heat isolation applied to the floor of the attic.

> **(Continued)**
>
> Modeling) enable better **collaboration of professionals and contractors** through the systems approach.[33]
>
> If there had been a common medium of communication and collaboration during the construction of the Monarga House, some unsolved problems, such as the water isolation and heat insulation systems in the attic, could have been solved.

Tectonic Affects Caused by Changes During the Application of Water Isolation

There is no tectonic affect created due to changes in the application of water isolation. This is because these changes are not visible.

Innovative Attitudes During the Application of Water Isolation

- This is the same problem which triggered an innovative attitude from one of the contractors during the tendering process. It relates to building a pool under the water tank and a boiler in the attic, during the construction of the roof. However, the interaction between the heat insulation and this pool was not sufficiently thought out and it reduced the effectiveness of the cool roof. This was a mistake made at this stage of construction.

> I talked to the builder about the work in the attic. He said he would put a 10 cm tall wall there to create a pool. I called the main controller about this, and he gave some detailed explanations to me and to the builder. The builder also said that the pipe to let the attic water out was too high and this might cause an accumulation of water. I called the contractor and asked him to ask the builder to solve this problem. That is the contribution of the builder.

Conclusion

Most of the changes to the electrical and mechanical systems disturbed the RONAC. The electrical engineer did not ask the RONAC about the positioning of the ACs while preparing the project. Similarly, the position of the electricity manhole was decided without telling her. The size of the water tank was

---

33 Francisco Valdes et al., "Applying Systems Modelling Approaches to Building Construction." In *33rd ISARC International Symposium on Automation and Robotics in Construction*. (Auburn, 2016): 844–852. DOI: 10.22260/ISARC2016/0102.

considered before the preparation of the mechanical engineering project but the requests of the RONAC were not considered in the project. The application project was shown to the RONAC; however, the process was so long that she checked the project too quickly and did not want to lose more time. Control of the application project solved many problems; however, many issues were missed as well. If these problems related to electrical and mechanical engineering projects are considered together with the changes in the civil engineering project (about increasing the reinforcement in the beams), it becomes clear that none of these engineering projects was fully applied in the Monarga House.

> *Reflection(s) from Theory*
>
> The philosopher Theodor W. Adorno asked for the simultaneous presence of **functionality and ornamentation in architectural details** to avoid the **fetishist characteristic of the functionless details**.[34]
>
> The leaf plaster and the water tank cannot be seen as both functional and aesthetic. However, there are some details in the Monarga House such as the two rain chains which are both functional and aesthetic.

Figure 5.8 presents the map of the tectonic process for Part B, which includes plastering and painting, the electrical and mechanical systems, ceramics and

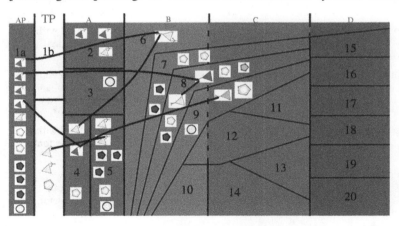

Figure 5.8 The map of tectonic process for Phase B - Application project (AP, 1a) and tendering process (TP, 1b), construction of foundations (2), the frame system (3), walls and openings (4), the roof (5), plastering and painting (6), the electrical system (7), the mechanical system (8), ceramics (9), water isolation (10), the windows, shutters, and front door (11), the interior doors (12), the wardrobes and cupboards (13), the fireplace and chimney (14), the garden walls (15), drainage (16), the garage and concrete work in the garden (17), metalwork (18), the pathways (19), and timber work in the garden (20).

---

34 Theodor W. Adorno "Functionalism Today" *Oppositions* 17, 1979: 31–41.

182 | *5 Tectonic Affects and Innovative Attitudes*

water isolation. This process is related to the previous phases, and the connections between the changes in different phases make this explicit. According to this map, all multiple changes originate from or relate to the application project and tendering process phases.

Table 5.5 presents the changes causing tectonic affects, the total number of changes (including the hidden changes), and innovative approaches during Phase B. The number of changes causing tectonic affects during the placement of mechanical systems is notable. Additionally, the greater number of innovative approaches within this phase is also noteworthy.

The actors who initiated the changes with tectonic affects within Phase B of construction of the Monarga House are seen in Table 5.6. The contractor, builder, foremen, and workers made seven changes with tectonic affects and the RONAC and the main controller made nine changes.

Table 5.5 The changes which cause tectonic affects and innovative attitudes during the Phase B.

Phase B	Eliminated changes	Singular changes	Multiple changes	Total number of tectonic changes	All changes	Innovative attitudes
Plastering, painting (B6)	None	None	1 positive	1 positive: 1	1	1 unrealized but caused positive tectonic affects.
Electrical system (B7)	None	2 positive, 2 negative	None	2 positive + 2 negative: 4	9	1 realized with no tectonic affects.
Mechanical systems (B8)	None	2 positive, 1 neutral, 1 negative	1 positive, 1 negative	3 positive + 1 neutral + 2 negative: 6	18	3 realized with no tectonic affects, 1 unrealized with tectonic affects.
Ceramics (B9)	1 positive	1 positive	1 medium	2 positive + 1 neutral: 3	7	1 realized with positive tectonic affects.
Water isolation (B10)	None	None	None	None	1	1 partially realized, but with no tectonic affects.
Total	1 positive	5 positive, 1 neutral, 3 negative	2 positive, 1 neutral, 1 negative	8 positive + 2 neutral + 4 negative: 14	36	5 realized but only 1 had tectonic affects, 1 partially realized, 2 unrealized and all caused tectonic affects.

Table 5.6 The actors initiating changes with tectonic affects in Phase B.

Phase B	Contractor/builder/worker	The RONAC/the main controller
Plastering/painting (6)	1 positive	1 positive
Electrical system (7)	1 negative	2 positive, 1 negative
Mechanical system (8)	2 positive, 2 negative	2 positive, 1 negative
Ceramics (9)	1 positive	1 positive, 1 neutral
Water isolation (10)	–	–
Total	4 positive + 3 negative: 7	6 positive + 1 neutral + 2 negative: 9

Note that some decisions were made collectively by all parties.

The reflections from theory in this chapter include various philosophies, architectural theories, tectonics and construction theories, and ethical theories.

The Reflections of Theory

- **Various philosophies** about time, such as linear, cyclical and fractal time, experience and authority relationships, the instrumental approach, care and authenticity, functionality and ornamentation of details and the fetishist character of functionless details, professionalization and the spontaneous ideology of architects.
- **Theories of construction** about CPM and PERT, building systems, coordination, collaboration, standards and building codes, being out of date, continuing the design while building, water problems in buildings, and the systems approach.
- **Theories of tectonics** about hidden components and elements, reading the logic of construction of components and elements, the playful character of timber, the articulation of surfaces which separate earth and sky, continuity with the natural context, and the tectonic importance of the visible parts of buildings.
- **Ethics theories** about the violation of agreement ethics, careful consideration of context, and corruption (fraud, poor-quality goods and services, reduced commitment to quality).
- **Theories of architecture** about not including workers, foremen, and technicians in the design process and the concept of camouflage.

6

Tectonic Affects and Innovative Attitudes due to Construction Changes Relating to Windows, Doors, Wardrobes and Cupboards, the Fireplace and Chimney

The windows, doors, shutters, wardrobes, and cupboards, and the fireplace and chimney (Part C) are important elements of design for the Monarga House. There were design ideas about the application of windows and shutters. These were about achieving maximum continuity between indoor and outdoor spaces. The fireplace also played an important role in the design of this house. It was designed to be at the center of the house. These issues had to be checked several times during the construction of the building.

Windows, Shutters, the Front Door – Changes, Tectonic Affects, and Innovative Attitudes

As opening the living area in four directions had a symbolic value and also contributed to the cross ventilation and climatic comfort of the living area, the character of these elements was important. Two of the three sliding window panels of the four living area windows were designed to be totally openable in order to provide continuity between indoor and outdoor spaces. Shutters also give character to the building as well as provide security. All aluminum windows and the front door have heat insulation and are double glazed. Shutters are lockable and their parts are movable. The building has a minimum number of openings to the west and north for minimum heat loss in winters and summers. The colors of these elements were also carefully selected.

Tectonics as a Process in Architecture, First Edition. Yonca Hurol.
© 2025 John Wiley & Sons, Inc. Published 2025 by John Wiley & Sons, Inc.

> *Reflection(s) from Theory*
>
> **Passive house design principles** for climatic comfort can be listed as continuous insulation, no thermal bridges, airtight layers, **high-performance windows and doors**, fresh air with heat recovery, shading, **orientation** and form, daylighting and solar gain, and moisture management. The application of these principles differs from climate to climate.[1] Many of these principles were considered in the design of the Monarga House.

Figures 6.1a and b show some of the openings in the building and the windowsills of the smaller windows. Figures 6.1c and d show a problem detail between the aluminum window rails and the floor ceramics. There was a gap between them, and this gap was closed with a metal sheet. Figures 6.1e–g depict the shutters of the house, the front door, the transom window over the front door, and the venting sidelight.

> At night I was worried about ending up with the wrong measurements. Because of that I called the builder to ask if he had different measurements for the four sliding doors in the living room or not. He said yes.

Changes During the Placement of the Aluminum Windows, Shutters, and Front Door[2]

Changes during this phase were as follows:

- The window heights were wrongly determined by the contractor, and this caused chaos regarding the heights of windows, doors, and cupboards and wardrobes. This problem has not been solved.

> *Reflection(s) from Theory*
>
> All **points and lines in architectural design are related to each other** and because of this they can achieve design principles which effect the perception of design elements and the affects they cause.[3]
>
> There are different door, window, cupboard, and wardrobe heights and the careless relationship between them in the Monarga House was contrary to the architectural design project.

---

1 Mary James and Bill James, *Passive House in Different Climates: The Path to Net Zero*. (Taylor and Francis, 2016).

2 Hashtags (#) indicate mistakes which could have been identified and solved by experienced foremen or technicians.

3 Mine Ozkar, *Rethinking Basic Design in Architectural Education – Foundations, Past and Future*. (Routledge, 2017).

Figure 6.1 (a) Windowsills in place. (b) Positioning of Window mullions. (c) The gap between the rails of the large windows and tiles. (d) The closure of the gap between the rails and ceramics with a metal piece. (e) Positioning of the aluminum window shutters and triangular roof openings. (f) The front door of the house with the wrong side panel. (g) The glazed venting sidelight of the front door.

- As mentioned earlier, the RONAC (Researcher, Owner, Neighbor, Architect, Controller) preferred sliding shutters, but later discovered that they cannot be locked, and their blades cannot be moved. So, she decided to use classic shutters. Before this she had also considered use of accordion doors and sliding shutters to be fully open in summer. However, because of the lack of climatic comfort caused by these types of windows and the bad quality of construction in North Cyprus, she gave up. The RONAC was also thinking about having a smooth floor level between the terraces and interior spaces. She gave this idea up to have good water isolation. The presence of window rails in that position would obstruct the continuity between inside and outside in any case.

Reflection(s) from Theory

To increase their interest, contractors may develop an **instrumental approach** towards buildings and decrease the floor and roof heights, minimize excavation, choose the same types of windows and shutters, reduce the number of rails in sliding windows and doors and always prefer similar details. This means that different parts, elements, and components of buildings are **enframed** to decrease the amounts of material used, decrease the quality of workmanship, and enable cheap workmanship.[4] Such *supplier driven models* create **monotonous and meaningless environments**.[5]

Since the construction of the Monarga House was carefully controlled, most changes applied to the enframed building parts, elements, and components were eliminated. However, the following sections of this chapter show that the dimensions of windows, interior doors, cupboards, and wardrobes of the Monarga House were determined with an instrumental approach.

- Construction of the sliding terrace doors began with two rails but was then changed to three rails. The project and the agreement specified three rails (#24). The elimination of this problem increased the continuity between the inside and outside because the openable area of windows increased.
- It became necessary to add a metal element between the large window rails and the ceramics to close the gap between them. This detail was not looking good and could have caused water problems in the future. The additional metal element contributes to the champagne color of the mullions in the interior space (#25). This was contributed by the main controller.
- It was planned that the aluminum front door with heat insulation would have an openable sidelight. However, as mentioned previously, the builder fitted a

---

4 Martin Heidegger, *"The Question Concerning Technology" In The Question Concerning Technology and Other Essays*, Trans. William Lovitt. (Garland Publishing, 1977[1954]a): 3–35. Martin Heidegger, "The Age of World Picture." In *The Question Concerning Technology and Other Essays*, Trans. William Lovitt. (Garland Publishing, 1977[1954]b): 115–154. Martin Heidegger, *"Science and Reflection" In The Question Concerning Technology and Other Essays*, Trans. William Lovitt. (Garland Publishing, 1977[1954]c): 155–182. Jacques Ellul, *The Technological Society*. Trans. J. Wilkinson. (Alfred A. Knopf and Random House, 1964). Jacques Ellul, *What I Believe*, Trans. G. W. Bromiley. (Eerdmans, 1989).
5 John Habraken, *The Structure of the Ordinary*. (MIT Press, 2000).

non-openable and opaque side panel. This panel had to be changed to one that was openable and transparent. Finally, a champagne color and a thin frame together with double semi-transparent glass were used for this sidelight. This created a nice color contrast with the front door and provided better illumination for the entrance area. Since the aluminum front door was shorter than the doorway, it became necessary to have a transom window above the door as well. This again increased the amount of light inside, but also created a security problem which was later solved with decorative metalwork.

> *Reflection(s) from Theory*
>
> **Minor architecture** goes **against authority formed through architectural mythologies.**[6] One way of achieving minor architecture in the Monarga House was to **mix different styles and colors** to avoid being identifiable and to be modest in order to avoid the myth of the architect as the subject. Minor architecture is different from a consistent application of one style. It is also different from **eclecticism**, which is usually highly impressive and takes up too much attention.
>
> The Monarga House combines different tastes and colors in a modest way. But the intention was not to have a crowded mixture of styles and tastes. The RONAC had to be careful about the **danger of having too many aesthetic styles simultaneously** in her house. One of the examples of this situation was elimination of fancy handles for the kitchen cupboards which did not go with the rest of the design.

- The RONAC thought about using stained glass for the windows of the front door and the bottom of the sliding door between the entrance area and the living room. This would separate her indoor and outdoor cats. However, she later decided to use frosted glass in these places and in the bathroom window and to use decorative metalwork for security purposes. Because of this, the idea of having stained glass was not realized.
- When they are open, the two small windows overlap with doors in the kitchen and the larger bedroom (#26). This problem was created by the contractor because he did not follow the architectural project and the others were not aware of this problem.

---

6 Jill Stoner, *Toward a Minor Architecture.* (MIT Press, 2012).

> **Reflection(s) from Theory**
> Just like points and lines, the **elements and components in buildings should also be related to each other** in architecture. Because of this, having clashing doors and windows is not acceptable for functional as well as aesthetic reasons.[7]

- The color of the shutters was changed to a lighter green to eliminate the dust problem. This was contributed by one of the RONAC's friends who deals with interior architecture.
- The shutter wheels had to be changed later because of their low quality. They were damaged within two years by the strong Cyprus sun. Later, the RONAC's son had to send more durable wheels.
- The color of the windowsills has been changed by the contractor. "The stone I have chosen for the windowsills had to be changed. It was too expensive, and I was not happy about my selection." The new one is more modest than the previously selected one.
- The sills of the triangular windows in the roof were forgotten. They were built of concrete. It was difficult to see and be aware of them, but they also protect the attic from water. This problem was solved later.
- There was a problem detail with the small windows and after a year it became impossible to close some of them. The metal pieces in all of them had to be changed to solve the problem. There was also a problem detail with the bathroom window involving its shutter when it was opened. This problem was also solved later.

> **Reflection(s) from Theory**
> **Care** is about **authenticity** which requires the **protection of the world and all beings,** of the possibilities of other humans and of one's own self because it is known that life is **temporary**.[8]
> There were some indications of **carelessness** in the construction of the Monarga House. Some of these are listed above concerning the detail problems of the windows, shutters, and front door. Although there were many other examples of carelessness in the production of the interior doors and fireplace, the foremen and technicians also did some careful work too.

---

7 Ozkar, *Rethinking Basic Design in Architectural Education.*
8 Martin Heidegger, *Being and Time,* Trans. John Macquarrie and Edward Robinson, 25th Edition. (Blackwell Publishing, 2005[1927]).

They also placed the triangular roof windows. These were really nicely designed. They were different from what I suggested. They used different and very useful aluminum profiles.

Tectonic Affects Caused as A Result of Changes in the Windows, Shutters, and Front Door

All of the aluminum work was done by one firm. This process took a long time and the RONAC thought the firm was too slow. However, the contractor's ignorance of the architectural project produced some important problems. There are also some improvements with respect to the architectural project (Table 6.1).

Reflection(s) from Theory

Continuity is poetic but the analytical mind separates things from each other. It is poetic imagery which relates things to each other for providing continuity.[9] Providing some continuity might be essential for achieving **meaning** in contemporary tectonics. **Separations** are usually made through **transitions** which avoid sudden changes in architecture. The Monarga House provides continuity between indoor and outdoor environments and the terraces provide a transition between them.

There can be various types of tectonic details. These are **tectonic system details** (to provide human scale, corners, places, and hiding places), **tectonic structural details** (such as Alvar Aalto's butterfly beams in Saynatsalo Town Hall) and **tectonic construction details** (such as Tadao Ando's window details in the Church of Light).[10] The transom and sidelight windows of the front door of the Monarga House, which also have metalwork, needed to be studied as a system detail as well as a set of construction details.

Innovative Attitudes During Placement of the Windows, Shutters, and Front Door

- The problem relating to the cancelation of the accordion windows and sliding shutters originated from the application project phase. This change could have triggered some innovative attitudes; however, it did not.

---

9 Anthony C., *Antoniades, Poetics of Architecture – Theory of Design*. (Van Nostrand Reinhold, 1992).

10 Yonca Hurol, *Tectonic Affects in Contemporary Architecture*. (Cambridge Scholars Publishing, 2022).

Table 6.1 Tectonic affects due to changes during the placement of the windows, shutters, and front door.

The change in 11	The symbol	The tectonic affect(s)	Responsible person/contributor	Related changes	Innovative attitude
Determining the wrong height for windows.		Caused chaos in respect of the height of the windows/doors and cupboards/wardrobes. Bad detailing.	Responsible person: the contractor	Many	None
Not having totally openable windows in the living area because of the window and shutter type and the difference between the floor levels inside and outside.		A decrease in tectonic continuity between the indoor and outdoor spaces.	Architect responsible from the application project, and aluminum firm staff warned the RONAC, and she took their advice about low-quality workmanship on the island.	Many	Unrealized and caused tectonic affects
Having three rails for the large aluminum windows instead of two.		This increased the opening surface of the windows and continuity between indoor and outdoor spaces.	Responsible person: the contractor, workers of the aluminum firm Solution: the RONAC	0	Unrealized and caused tectonic affects
Having an additional metal element between the aluminum windows and the floor ceramics.		Provided a more tectonic detail and eliminated a future water problem.	Responsible people: aluminum firm Solution from: the main controller	0	Realized and caused tectonic affects

Having semi-transparent transom and sidelight windows attached to the front door.		Provided a more tectonic detail and better illumination.	Responsible people: aluminum firm Solution from: the RONAC	Many	Realized and caused tectonic affects
Color of the shutters was changed to a lighter green to eliminate dust.		Provided a more tectonic outlook with care because this solves the dust problem on shutters.	Solution: an architect friend of the RONAC	0	None
Two windows overlap with doors when they are opened.		Provided a bad detail because this is against architectural knowledge.	Problem: the contractor	0	None
Color of the windowsills has been changed.		This creates tectonic harmony with the floor edge marble pieces.	The contractor	0	None

Eliminated changes/problems are shaded in light gray in the table. Changes which cause another change or originate from another change are shaded in dark gray. Changes that did not cause any other changes are unshaded. Symbols reflect not only the type of change, but also associated feelings. Lighter colors in symbols reflect positive feelings.

- The RONAC did not agree with the builder's idea of substituting three-railed windows for two-railed ones, so they were canceled because of the negative tectonic affects they caused. However, this situation could have triggered some other innovative attitudes resulting in better solutions.
- Closing the gap between the ceramics and the aluminum window rails with an aluminum surface was the idea of the main controller and the result of an innovative attitude.
- Changing the side panel to the front door as an aluminum venting sidelight and having the transom window above the front door were results of innovative attitudes. The first was the idea of the RONAC and the second was the solution of the builder.
- Although the triangular windows in the roof were examined in the section about the roof, the aluminum part of these windows was designed by the builder who dealt with the windows and shutters. The aluminum part of these triangular windows included an anti-insect net as well as specially designed aluminum pieces to allow air to enter but not water. This was a result of an innovative attitude, and it worked well.

Reflection(s) from Theory

Some building details are specially designed for that particular building and they have high tectonic value. The butterfly beams of Alvar Aalto are a good example of **specially designed tectonic details**. There can also be **typical details**, found on the Internet and applied to many buildings. Aluminum mullion details are examples of typical details. It is also possible to **adapt details** to different situations.[11] Some typical details should be redesigned and adapted for the building for many reasons. System details are usually adapted details. Designers study photographs and detailed drawings of various buildings and design new details in order to catch the image in their minds. The triangular window detail on the roof of the Monarga House was adapted from the detail of other roof windows which have different forms.

They found a dead baby bird in the attic. Somehow it died in my house. Either it got stuck there, or it was ill and entered my attic before it was fully closed and died there ... Sad in any case.

---

11 Hurol, *Tectonic Affects in Contemporary Architecture*.

Interior Doors – Changes, Tectonic Affects, and Innovative Attitudes

The interior doors of the house were designed as standard fabricated doors. However, as the heat insulation on the reinforced concrete frame elements was canceled, the RONAC and the main controller negotiated with the contractor to have timber frames for interior doors. This added to the presence of timber in the house and more timber elements were added later. The sliding door between the entrance hall and the living area was intended to be an aluminum sliding door. However, since the ceramics had been placed and aluminum sliding doors needed a base, its application became impossible. The contractor suggested having a timber sliding door, which can be hung from the top, and the RONAC accepted this. This also contributed to the presence of timber inside the house. Figures 6.2a–c show the frame of one of the interior doors, a completed interior door, and the sliding door in the entrance hall.

> *Reflection(s) from Theory*
>
> **Compensation for mistakes** during construction can be a part of **change management** which is usually effected through **change orders.**[12] However, compensation for mistakes was made through face-to-face agreements during the construction of the Monarga House. To compensate for the mistake in relation to the application of heat insulation on the structural elements, the factory-made frames of the interior doors were substituted by handmade timber ones. To compensate for the detail mistake in relation to the entrance hall sliding aluminum door, the contractor suggested having a timber sliding door.

Changes During the Placement of Interior Doors[13]

- Having the interior door frames (but not leaves) in timber was a change. Industrial doors were specified in the original project. This was a great contribution to the tectonics of the house. Having timber door frames together with timber pergolas was an inspired choice. Later, timber was used as a belt

---

12 Alia Alaryan et al., "Causes and Effect of Change Orders on Construction Projects in Kuwait" *Journal of Engineering Research and Applications* 4, no. 7, 2014: 1–8. Qi Hao et al., *"Change Management in Construction Projects" International Conference on Information Technology in Construction CIB W78 2008.* (Chile: Santiago, 2008).

13 Hashtags (#) indicate mistakes which could have been identified and solved by experienced foremen or carpenters.

Figure 6.2 (a) Timber interior door frames in place. (b) The doors installed. (c) The sliding timber door in the entrance hall.

on the fireplace, and in various parts of the garden. This contributed to the richness of natural materials in the house. However, the frames and leaves of the interior doors had to be produced by different carpenters, firms, and builders. "There might be color problems between the frames and doors. I am worried. But I asked them to show me before making them."

> *Reflection(s) from Theory*
> **Natural building materials**, such as stone, timber, and adobe, are **less homogenous** than modern building materials, such as reinforced concrete, steel, and laminated timber. If they are used as the structural material, this feature makes them weaker than modern structural materials. Whether they are used as structural material or for construction components, natural materials usually require **manual labor**. Their non-homogenous features and the use of manual labor make the elements and components produced with natural materials **more haptic** and **less ordered** compared with modern materials. These features make these elements and components closer to nature. They are from nature. One of the strongest ways of achieving **tectonic continuity with nature** is to use natural materials.[14]

- The heights of the timber door frames were much lower than those of the aluminum windows. Still, there had to be transom windows at the top of these door frames because they were too tall as timber door frames (#27). These transom windows did not feature in the original project. However, the RONAC was happy

---

14 Hurol, *Tectonic Affects in Contemporary Architecture*.

Interior Doors – Changes, Tectonic Affects, and Innovative Attitudes | **197**

with them. The mistake in relation to the height of windows and interior doors was made by the contractor when placing the lintels.

- The cornices of the interior doors were installed by the firm that produced the door leaves. The cornices of two doors did not cover the gap between the door frame and the wall. The gap was large (#28). The firm suggested using gypsum to close these gaps. Since this could have caused cracks later, the RONAC did not want this and the firm had to replace those cornices with larger ones. They said the largest cornice size was 9 cm and they could not find any others. The RONAC asked them to solve this problem. They used 10-cm wide cornices. Since this increases the amount of timber in the house, it contributes to the tectonics of the house.

Reflection(s) from Theory

According to Neil Leach, a desire for **camouflage** in architecture indicates a desire to be connected.[15] Camouflage transforms parts of the building, making them look better, and this results in a better connection with other people due to having a better image. However, camouflage with gypsum as suggested for the Monarga House, cannot last long. Therefore, one should also consider the **temporariness of camouflage**. On the other hand, camouflaging the differences in the height of windows, interior doors, cupboards, and wardrobes with the help of ceramics is not temporary camouflage.

- Some cornices were colored green (#29). The RONAC asked for them to be changed. This was accepted.
- Mastic was not applied to the cornices, but finally the firm had it done because of the consistent demand of the main controller and the RONAC (#30). The inside door stoppers were forgotten but another carpenter provided them as a present for the RONAC. Problems related to the interior doors caused several changes.
- Having a timber sliding door instead of an aluminum one contributed to the presence of timber in the house. Only the pergolas and the headboard of the roof were timber in the original project. However, the sliding timber door became a problem because the carpenter used the wrong timber for some pieces, then painted the whole door the same color, and that color did not match the color of the timber in the house. The RONAC had to accept this color. The door handle of the sliding door was carved by the carpenter because the handles available on the market were not appropriate. There were also several changes caused by the sliding door, and the carpenter later informed the RONAC that he was never paid by the contractor for this door.

---

15 Neil Leach, _Camouflage_. (MIT Press, 2006).

> **Reflection(s) from Theory**
> According to Albert Chan and Emmanuel Owusu, forms of **corruption** in the construction industry include **deception** which may include not paying the carpenter.[16]

- The timber trap door which gives access to the attic was designed by another carpenter. The original project specified an automatic ladder with its own panel. However, those ladders cannot be found on the market for ceilings higher than 3 m. Therefore, the solution had to be redesigned. The use of timber for this trap door also contributes to the presence of timber in and around the house.

> **Reflection(s) from Theory**
> **The repetitive use** of some design elements in a building helps to create **visual cohesion** and reinforce the design principles of **hierarchy, unity,** and **rhythm**.[17] The use of timber in various parts of the Monarga House contributed **hapticity** as well as **unity** and **rhythm,** besides providing **tectonic continuity with nature.**

Tectonic Affects Caused due to the Changes in the Processes of Interior Doors (Table 6.2)

The timber door frames were built by a carpenter and the leaves of the doors were produced by another firm. The RONAC preferred to keep the natural greyish beige color of paste on these door leaves. "They thought nobody had warned me about the marks of the press on the doors. I told them that I had been warned. The firm's architect said that when the polish is put on, some press marks will get darker. Still, I decided to go for the natural color of paste on the door. I still think it is lovely. It is like a greyish cream, and it looks soft ... I am very happy about selecting this natural color of the paste."

---

16 Albert P.C. Chan and Emmanuel K. Owusu "Corruption Forms in the Construction Industry Literature Review" *Journal of Construction and Engineering Management* 143, no. 8, 2017: 04017057.
17 Jennifer Gaskin, *A Brief Guide to Repetition – A Design Principle, Venngage Blog,* 2022, accessed February 22, 2025. https://venngage.com/blog/design-principle-repetition/#Repetition%20Design%20Principle%20FAQ.

Table 6.2 Tectonic affects due to changes during the montage of the interior doors.

The change in 12	The symbol	The tectonic affect(s)	Responsible person/contributor	Related changes	Innovative attitude
Having the interior door frames in timber.		Use of natural material, haptic tectonic affect.	Suggested by the main controller, and accepted by the contractor.	Many	None
Having frames and leaves of interior doors different from each other.		Unusual tectonic detail.	The RONAC	0	None
Having the height of interior doors lower than that of the windows.		Bad tectonic detail.	Contractor	Many	None
Having small transom windows over the interior door frames.		Unusual tectonic detail.	Carpenter	0	Realized and caused tectonic affects.
Eliminating the wrong cornice details.		Eliminated wrong tectonic detail.	Problem: the firm making the door leaves, wardrobes, cupboards, etc. Solution: the RONAC and main controller.	0	Realized and caused tectonic affects.

(Continued)

Table 6.2 (Continued)

The change in 12	The symbol	The tectonic affect(s)	Responsible person/contributor	Related changes	Innovative attitude
Having a timber sliding door.		Use of natural material, haptic tectonic affect, contribution to the presence of timber in and around the house.	Suggested by the contractor, and accepted by the RONAC.	0	None
Having a timber trap door to access the attic.		Use of natural material, haptic tectonic affect, contribution to the presence of timber in and around the house.	The RONAC's decision.	Many	None

Eliminated changes/problems are shaded in light gray in the table. Changes which cause another change or originate from another change are shaded in dark gray. Changes that did not cause any other changes are unshaded. Symbols reflect not only the type of change, but also the associated feelings. Lighter colors in symbols reflect positive feelings.

Interior Doors – Changes, Tectonic Affects, and Innovative Attitudes | **201**

> *Reflection(s) from Theory*
>
> Keeping the natural color of materials and building elements and components can be explained through theories of tectonics. According to Eduard Sekler, **concealing** a building part is an **atectonic** approach.[18] Similarly, according to Kenneth Frampton, the **logic of construction of building parts should be readable** in order to be accepted as **ontological**.[19] Therefore, keeping the natural color of the paste on the interior doors and not painting them can be seen as a tectonic and ontological approach. However, not painting the door might cause some problems in the long term. Even brutal concrete façades of buildings are covered with protective paint. The critical point about painting is the intention behind it. If it is for protection, this is quite different from hiding or faking something. For example, painting the concrete applied to the roof tiles red, to cover the gaps between them, makes concrete look like red tiles. The RONAC did not accept this atectonic approach for the roof of the Monarga House.

Production of the timber sliding door by the same carpenter who produced the timber door frames was slow. The RONAC believes that this delay was due to payment problems caused by the contractor.

Most of the tectonic affects due to changes during the construction of the interior doors cause positive feelings. This is mainly because of the use of timber. However, there were negative tectonic affects due to the difference between the heights of the doors and windows; this was because of the contractor's ignorance of the architectural project. There were seven such changes causing tectonic affects during this stage.

> The carpenter was worried about the way the glass was attached to my doors (with silicon) … We solved this problem by using thin pieces of timber at the connection points.

---

18 Eduard Sekler, "Structure, Construction, Tectonics." In *Structure in Art and Science,* Ed: G. Kepes. (1965): 89–95, accessed August 18, 2018. https://610f13.files.wordpress.com/2013/10/sekler_structure-construction-tectonics.pdf.
19 Brian Kenneth Frampton, "Rappel a l'Ordre: The Case for the Tectonic." In *Labour, Work and Architecture,* Ed: K. Frampton. (Phaidon, 2002): 91–103. Brian Kenneth Frampton, *Studies in Tectonic Culture – The Poetics of Construction in Nineteenth and Twentieth Century Architecture.* (MIT Press, 1995).

> *Reflection(s) from Theory*
>
> **Hapticity** is caused by a simultaneous effect on the five senses. It is not only natural materials which have haptic characteristics. The silky, smooth, and cool surfaces of metals as well as the rough gray surfaces of brutalist concrete also have haptic characteristics. The domination of the visual sense over the other senses in respect of contemporary architecture is a real-life problem, which inspired Juhani Pallasmaa to conduct research into hapticity. He described the role of the five senses and also articulated the effects of light, shadow, and detail errors in relation to hapticity.[20] Hapticity is one of the **poetic tectonic affects** in architecture.[21]
>
> The use of timber for the interior door frames and the sliding door in the entrance hall contributed to the hapticity of the Monarga House.

Innovative Attitudes During the Placement of Interior Doors

- It became necessary to add transom windows above the timber frames of the interior doors because the door height was beyond the limit of timber frames. To solve this problem, the carpenter developed an innovative attitude and added another timber element and a transom window to the top of each interior door.
- Since the gaps between some door frames and their lintels were large, the building firm suggested covering the gaps with gypsum which is not a good idea. This triggered an innovative attitude and both the RONAC and the main controller demanded larger cornices and the problem was solved in a successful way. This idea eliminated some negative tectonic affects due to later deterioration.

> *Reflection(s) from Theory*
>
> Even if the intention in preparing the application project is to design all details of the building, there is always some **ambiguity** remaining. This ambiguity occurs due to many factors, such as finding a better material than the one specified in the project. Therefore, it becomes necessary to **design some details during the construction** of buildings. Such changes sometimes cause serious differences between the application project and the completed building.[22]

---

20 Juhani Pallasmaa, *The Eyes of the Skin – Architecture and the Senses.* (John Wiley, 2005).

21 Gaston Bachelard, *The Poetics of Space*, Trans. M. Jolas. (Beacon Press, 1994[1958]). Hurol, *Tectonic Affects in Contemporary Architecture.*

22 Mahairi McVicar, *Precision in Architecture – Certainty, Ambiguity and Deviation.* (Routledge, 2019).

Cupboards and Wardrobes – Changes, Tectonic Affects, and Innovative Attitudes

The RONAC preferred to have a modern look for the wardrobes and cupboards in order to break the traditional character of the house. This idea contributed to the concept of minor architecture which was achieved through mixing different ideas in order to avoid the dominance of one style. Mixing different approaches creates a more modest atmosphere.

> *Reflection(s) from Theory*
>
> The anthropologist Tim Ingold compared buildings with hills and wrote that hills are **close to all beings** and birds; insects as well as humans can approach them without any concern. However, this is not true for buildings. Buildings keep many beings, including some humans, away from them.[23] This may be due to some security measures or by the psychological messages given by the building.[24] If we consider the philosopher Martin Heidegger's definition of space as **clearing out**, we may also recognize that **being distant** is one of the common features of architecture. Clearing out is done for something to emerge there.[25] However, there can be more modest and elite approaches to architecture and the Monarga House was designed to be modest.

The firm, which also repaired the cornices of the interior door frames and their leaves, produced the cupboards and wardrobes. Figures 6.3a and b demonstrate one of the wardrobes during its positioning and montage, and some cupboards in the kitchen. Figures 6.3c and d show the main bedroom wardrobe and the kitchen sink.

> They brought the doors of cupboards for the kitchen and wardrobes in the bedrooms. The color and material of their doors were lovely. I was happy about my selection. Then I had answers to some of my questions. They were producing good work.

---

23 Tim Ingold, *Making – Anthropology, Archeology, Art and Architecture*. (Routledge, 2013).
24 Mike Davis, *City of Quartz – Excavating the Future of Los Angeles*. (Verso Books, 1990).
25 Heidegger, *Being and Time*.

6 Tectonic Affects and Innovative Attitudes due to Construction Changes

Figure 6.3 (a) Parts of wardrobes and cupboards. (b) The completed wardrobe and cupboard montage. (c) The gap at the top of the wardrobe in the main bedroom. (d) Montage of the kitchen counter tops and the kitchen sink which are nearly completed.

Changes During the Positioning of Wardrobes and Cupboards[2]

So, we agreed to keep them at the same level as the windows or doors … But now I am not happy with this. I talked to them honestly. I do not know what the solution is. They created a third level in my interior spaces: the heights of the doors, windows, and wardrobes are different from each other … During this conversation I understood that the whole problem is due to the cornices around these wardrobes. The length of their cornices is limited to 2.78 meters, and they have trouble if the wardrobes are taller than that. If that is the case, there occurs a little problem detail which has to be covered with paste. They are worried about that.

- The same problem caused by the height of the doors occurred with the height of the wardrobes and cupboards. Although the RONAC told the firm that they must be the same height as either the interior doors or the aluminum windows, the firm produced all of them with different heights. The kitchen cupboards are low, but the wardrobes are high (#31). This causes an aesthetic problem in the interior spaces, because these items do not refer to each other. This problem was partially solved in the kitchen and bathroom with the help of a ceramic organization.
- These height differences caused some other problems too. One of these is the gap at the top of the main bedroom wardrobe (#32). The firm again suggested covering this gap with gypsum, but the RONAC and the main controller did not accept this. Instead, a colored piece was added to the back of the gap and that place became a niche. The same gap also occurred at the top of the second bedroom wardrobe, but this gap was larger. As a solution two wardrobe doors were added to the front of this gap.

Reflection(s) from Theory

Camouflage covers something, hides it, and changes its image in an atectonic way.[26] To avoid **fetishism** in architecture, Theodor W. Adorno suggested architects should **combine functional and aesthetic features in details**.[27] Covering the gap at the top of a wardrobe in the Monarga House with gypsum was an attempt to camouflage; however, it was also a **temporary camouflage**. Adding more wardrobe doors to the front of the gap at the top of another wardrobe was a **functional solution** which also makes the wardrobe look aesthetically better. Transforming the gap on top of another wardrobe into a niche, which is not useful, was a somewhat **aesthetic detail**.

- The kitchen cupboards were much lower than the doors and windows in the kitchen. This was different from the original project. There were three different levels in the kitchen and this problem was getting complicated because of the presence of the ceramics on the kitchen walls. This problem was solved to a certain extent by organizing the ceramics on the walls. The height of the kitchen wall ceramics matched the height of the windows, which were higher than the doors and cupboards.
- One cupboard shelf was placed against one of the electric connections inside the cupboard (#33). The firm changed the placement of the shelf.

---

26 Leach, *Camouflage*.
27 Theodor W. Adorno "Functionalism Today" *Oppositions* 17, 1979: 31–41.

> *Reflection(s) from Theory*
>
> There can be several **elements, components, and details hidden** in buildings. Structural elements can be hidden. Le Corbusier's Chapel of Notre Dame de Haut at Ronchamp is a successful example of this, because its vertical structural elements are covered.[28] Construction components can be hidden. A double skin façade can be an example of this situation. Details can also be hidden. The inner layers of pitched roofs are hidden whether they are done correctly or not. The electrical connection behind the cupboard shelf was left like that because it was a **hidden error**. The RONAC had it corrected. Eduard Sekler's **atectonic**[29] concept which is about hidden items in buildings needs adapting to hidden errors in new tectonics theories.

- The wrong cupboard door handles were brought, but the architect of the firm realized this in time, and they were changed. This mistake could have ruined the character of the kitchen because the wrong handles were very elaborate.
- The original project specified a cupboard for the gas cylinder in the garden. This was not realized. Instead, the gas tube was covered with a piece of green cloth to protect it from the sun.
- The firm planned to cover the top of the oven with stone as an addition to the kitchen counter top. However, the type of oven was different, and this was impossible. This problem was identified on time by the technician of the firm while he was measuring the placement of the kitchen counter top (#34).
- A kitchen tap was placed in front of the window mullion to avoid it being seen from outside. This was another contribution of the same technician (#35).

Tectonic Affects Caused due to Changes in the Production and Placement of Wardrobes and Cupboards

Tectonic affects due to changes in the production and montage processes of wardrobes and cupboards were mostly problematic because of a carelessness of the firm which produced and installed them. This was the same firm which produced the leaves and cornices of the interior doors. There were five changes causing tectonic affects due to this process.

---

28 Hurol, *Tectonic Affects in Contemporary Architecture.*
29 Sekler, *"Structure, Construction, Tectonics."*

> *Reflection(s) from Theory*
>
> Three ways for the **application of power** are discrimination, hegemony, and ignorance. Discrimination is against human rights.[30] For example, if the building quality differs according to the identity of the owner, this can be called discrimination. **Hegemony** occurs if a person or a group of people dominate others.[31] It may be possible to talk about the dominance of builders, foremen, and workers by professionals in the contemporary building industry and this is the accepted procedure. However, if the builders, foremen, and workers dominate or somebody who is not authorized dominates, that can be called hegemony. **Ignorance** happens when people do not take action to eliminate crime, hegemony, or discrimination, and so on. It may be difficult for those who have been ignored to talk about their problems, because the identification of ignorance is not easy.[32] If a controller does not report problems to the owner, this can be accepted as ignorance.
>
> There is ignorance about the production and placement of cupboards and wardrobes in the Monarga House. The demands of the RONAC about dimensions were ignored by the firm. The RONAC also realized the situation after the cupboards and wardrobes were installed and she could not demand anything. This is because any request to make changes was going to be costly as well as cause serious delays in the construction (Table 6.3).

I told them that the key was in the electric box. Later, I worried that they might have brought the wrong handles again. I called the architect of the firm ... After exercising, I rushed home to see if the handles were the correct ones or not. There was a thunderstorm. I entered the house in darkness and found one of the handles with my hands and thought that they were OK. I also saw a little bit of the color in the dark. So, I was sure about them. I took photos of them the following morning.

---

30 United Nations, *Human Rights*, accessed February 16, 2023. https://www.un.org/en/global-issues/human-rights.

31 Ernesto Laclau and Chantal Mouffe, *Hegemony and Social Strategy: Towards a Radical Democratic Politics.* (London: Verso, 1985).

32 Gayatri Chakravorty Spivak, *Can the Subaltern Speak? Reflections on the History of an Idea.* (Colombia University Press, 2010[1985]).

6 Tectonic Affects and Innovative Attitudes due to Construction Changes

Table 6.3 Tectonic affects due to changes during the placement of wardrobes and cupboards.

The change in 13	The symbol	The tectonic affect(s)	Responsible person/ contributor	Related changes	Innovative attitude
Having the height of wardrobes and cupboards different from that of the windows and doors.		A bad tectonic detail which is partially solved in the kitchen and bathroom and special designs were applied for the bedrooms.	The firm which produced the wardrobes and cupboards and the contractor.	Many	None
Solution to the gaps above the bedroom wardrobes.		Two different tectonic details were adopted for those gaps.	Responsible people: the firm which produced the wardrobes and cupboards. Solution: the RONAC	Many	Realized and caused tectonic affects.
The wrong kitchen cupboard handles were changed.		A bad tectonic detail creating disharmony in the kitchen was eliminated.	Responsible people: the firm which produced the wardrobes and cupboards. Solution: the architect of the same firm	0	None
The cupboard for the gas cylinder was forgotten.		A bad tectonic detail.	Unknown – ignored by everybody.	0	None
The kitchen tap was placed in front of the window mullion to avoid it being seen from outside.		A good tectonic detail.	Solution: a technician of the firm.	0	None

Eliminated changes/problems are shaded in light gray in the table. Changes which cause another change or originate from another change are shaded in dark gray. Changes that did not cause any other changes are unshaded. Symbols reflect not only the type of change, but also the associated feelings. Lighter colors in symbols reflect positive feelings.

Innovative Attitudes During the Placement of Wardrobes and Cupboards

- After the agreed height of wardrobes and cupboards was changed by the building firm some strange spaces occurred at the top of all wardrobes and cupboards. The building firm suggested covering these spaces with gypsum, which is not a good solution. This triggered an innovative attitude and the RONAC and the main controller suggested other solutions for each wardrobe and cupboard.

The Fireplace and Chimney – Changes, Tectonic Affects, and Innovative Attitudes

The fireplace has an important role in the design of the Monarga House. The living area of the house was designed to be continuous with the outdoor spaces by opening in four directions and to have a fireplace in the middle of this space. The character of this fireplace was also changed several times during the design and construction process of the house. Finally, just like the wardrobes and cupboards, it also became a modern fireplace.

The design of the chimney, its height, and the chimney cap also changed several times during the design and construction processes of the building. Its first design was like the chimney of a house which inspired the RONAC a lot. This is the house in Figure 3.4a, but the photo does not display its small chimney. However, since the chimney of the Monarga House is much larger, the RONAC had to change her idea. She once designed a copper-like chimney cap, resembling a witch's hat. Later, she decided to have a simple four-sided pitched tile cap for this tall chimney.

Reflection(s) from Theory

The **symbol of the witch** represents **female empowerment** within contemporary patriarchal societies.[33] The RONAC's idea of having a chimney cap in the form of a witch's hat was not a result of rational thinking. She thought this would fit with the idea of the house and it would also look good.

Figures 6.4a and b demonstrate some phases of the construction of the modern fireplace. Figures 6.4c–e show the final form of the chimney and the chimney cap, the fireplace with a timber belt around it and its painted form with a special paint which shines like copper under light.

---

33 Sophia Quaglia, "Women are Invoking the Witch to Find their Power in a Patriarchal Society" *Quartz*, 2019, accessed February 22, 2025. https://qz.com/1739043/the-resurgence-of-the-witch-as-a-symbol-of-feminist-empowerment.

210 | *6 Tectonic Affects and Innovative Attitudes due to Construction Changes*

Figure 6.4 (a) Montage of the two-sided modern fireplace. (b): The fireplace insulated. (c) The chimney top with its four-sided cap covered by red tiles. (d) The timber belt round the fireplace. (e) The painting of the fireplace.

Reflection(s) from Theory

Since the color of **copper** evokes fire, the RONAC wanted to paint the fireplace a copper color and also to have a copper chimney cap. Copper can also be used as a covering material for roofs. The name copper originates from the Latin word **cuprum** and cuprum is related to **cyprium** which means of **Cyprus**.[34]

Then they put on the glass fireplace doors, and we made a fire. It became hot and burned very quickly. They made the place dirty, but they cleaned it before leaving. Then they left but they had to come back to close the top of the chimney.... This took the whole day, and I am tired now but happy. The heart of my house is there now. After this, I locked the door and kept the key because the main controller told me that some people have parties in houses under construction if there is a fireplace.

---

34 Madeleine Hunter, *"Cyprus and Copper" NCM New Cyprus Magazine*, accessed February 16, 2023. //newcyprusmagazine.com/cyprus-and-copper/.

The Fireplace and Chimney – Changes, Tectonic Affects, and Innovative Attitudes | **211**

> *Reflection(s) from Theory*
>
> The philosopher Gaston Bachelard wrote about the **psychoanalysis of fire** and related fire to the Prometheus and Empedocles complexes which originate from Ancient Greek myths. The Prometheus complex explains why it is so important and ethical for humans to challenge existing rules in order to develop better ones and to foster innovation. The Empedocles complex reflects the rather negative psychological tendency of humans to burn out like a piece of wood.[35] Both complexes show the meaning of fire for humans.
>
> Gottfried Semper used the Caribbean House as a model to define the four elements of architecture which combine technical issues with life issues related to architecture. These four elements are the earthwork (such as foundations), the **hearth** (the fireplace), the framework and roof (the lightweight superstructure made of natural materials), and the enclosing weave (the handmade covering surfaces). Here the hearth appears as the main gathering place (for cooking, eating, and negotiating) for users of the house. It is in the middle of the house.[36]
>
> The fire and fireplace have an important place at the Monarga House. The fireplace is at the center of the house and gives character to the living area and the whole house.

Changes During the Construction of the Fireplace and Chimney

- The RONAC realized during construction that there was a serious problem with the positioning of the fireplace in the living area. She designed it to sit between two interior doors and the idea was to have four parts to the living area. However, the position of the fireplace was moved forward during the application project phase. Its new position was almost between the two sliding windows, and this went against the preliminary design project. It was going to block cross ventilation as well as the view. It was also causing a problem for the study area in the living room. The RONAC remembers that the architect who dealt with the application project was trying to enlarge the dining area. He thought that in the future somebody would like to put a large dining table there. The RONAC spoke to the owner of the fireplace firm, and it was decided to move it as far back as possible. The flue opening of the chimney is at the front, but the body of the fireplace was moved backwards, and this solved the problem.

---

35 Bachelard, *The Poetics of Space*.
36 Gottfried Semper, *The Four Elements of Architecture and Other Writings*, Trans. Harry F. Mallgrave and Wolfgang Herrmann. 2nd Edition. (Cambridge University Press, 2010[1851]). Gottfried Semper, *Style in the Technical and Tectonic Arts or Practical Aesthetics*, Trans. Harry F. Mallgrave. (Getty Research Institute, 2004[1860]).

> **Reflection(s) from Theory**
>
> Some of the changes made during the application project phase were done to **increase the commodity value**[37] of the Monarga House. The original project was designed only for the RONAC. The RONAC thought that it was reasonable to consider the future users of the house and make some changes without disturbing the RONAC's needs.

- The owner of the fireplace firm warned the RONAC that the fireplace in the project was too large, which could restrict the circulation areas around it (#36). The RONAC accepted this and a smaller version of the same type of fireplace was fitted. "Today I received an email from the fireplace firm. They suggested a smaller fireplace. I asked them why. I also asked them to build a classical type of fireplace with a top. Later, I found out they were worried about the space adjacent to two sides of the fireplace. Also, it was not possible to have a top for these industrialized fireplaces."
- The fireplace was two directional in the preliminary design project, but it was one directional in the application project. However, the contractor accepted that it was to be two directional while the contract was being prepared.
- There was a fireplace top in the original project, but with a modern fireplace this was not possible. Because of this, a timber belt was made to replace the top. Having a timber belt contributed to the presence of timber in the house. This was a contribution of one of the RONAC's architect friends. However, the belt had four pieces on its front face. So, the RONAC refused it. Then the contractor called her to ask whether this problem could be solved by adding cornices on all faces of the belt. The RONAC accepted that. "When I called the carpenter for the last time, he said people from the heat insulation firm came. He opened one of the shutters for them. He was worried about his timber work on the fireplace. He said they put nylon around the fireplace. I have to go and see it today. I was also worried."
- A special brown (with a copper hue) color was selected for the fireplace with the help of a painting firm and a friend of the RONAC.
- The chimney was built lower than it appeared in the project, but its height was increased later (#37). It was below the level of the roof parts of the building, but currently it is 60 cm above the highest level of the roof. "The height of the chimney was raised. It looks big and beautiful as the heart of the house. I have to design a cap for this chimney. Because it is thick and covered with leaf plaster."

---

37 Judith O'Callaghan "Architecture as Commodity, Architects as Cultural Intermediaries – A Case Study" *Architecture and Culture* 5, no. 2, 2017: 221–240.

The Fireplace and Chimney – Changes, Tectonic Affects, and Innovative Attitudes | **213**

- The chimney cap was designed differently in the original project and redesigned several times during the construction. However, since this is a large chimney, aesthetically it was more suitable to have a simple four-sided metal chimney cap. The RONAC also did not want to cause any danger from lightning because of having a sharp chimney cap in metal. " I decided to have a copper chimney … On the way back home, I saw that the tiles for the chimney cap had been placed. I became very excited again … These painters had also painted the top part of the chimney darker. I loved it."
- Since the chimney was closed later, the main problem was the water coming in via the chimney flue. If water comes in and moistens the heat insulation material, this can harm the heat chamber of this type of modern fireplace.

Reflection(s) from Theory

If we consider Gottfried Semper's categories of tectonic (timber or steel frames) and stereotomic (a stone, brick, or adobe masonry structure or reinforced concrete frame),[38] **old fireplaces have stereotomic structures** made out of stone, brick, or both. However, **modern fireplaces have tectonic structures** made of steel frames. The use of space also differs within these fireplaces. The traditional fireplace heats via its mouth and walls and has some details inside to suck the smoke upwards. However, modern fireplaces hide a metal body inside and there is space around it. There is also space at the top of the metal body, and it is called the heat chamber. The metal pipe of the fireplace passes through the heat chamber which also has grilled openings to give out heat and heat the spaces around it. Traditional fireplaces have one stereotomic chimney. However, modern fireplaces have a hidden metal pipe inside the stereotomic chimney. This means that the **tectonic character of the modern fireplaces is radically different from** that of traditional fireplaces.

- The contractor had a very practical tip about stopping birds getting inside the chimney. He said he did the same thing in his house (#38). He cut a bird cage and attached it to the inner surfaces of the chimney, and this stopped birds for a long time. We had a similar solution to this problem.

---

38 Semper, *The Four Elements of Architecture and Other Writings; Semper, Style in the Technical and Tectonic Arts or, Practical Aesthetics.*

> *Reflection(s) from Theory*
> Taking architectural precautions to keep birds away from chimneys, ACs, and terraces are typical architectural techniques for **clearing out the space** as Martin Heidegger suggested.[39] This shows the limits of the Monarga House having **continuity with nature**.[40] The Monarga House is not close to all beings like a hill.[41]

Today there was a thunderstorm, and I saw from my window that the temporary nylon cover at the top of my chimney had blown off. I was worried first about damage caused by the bricks on top of it (because they fell onto the roof tiles). But later I saw there was not very serious damage. However, the main problem was water coming in from the chimney ... This might lower the capacity of the fireplace. When I visited the house, I saw there was no water inside the fireplace. This means there was not much water in the metal flue inside the chimney. Since the size of the opening is almost equal to the metal chimney, there cannot be much water getting into the insulation material either. Tonight, I will pray for a dry night and call the contractor tomorrow to ask him to send somebody to cover the chimney again. I left a nylon cloth and an cable inside the house for them to cover the chimney tomorrow.

Since it was also raining, I started to worry. I called the contractor several times, but they were late as usual. I had to push them to cover the chimney. I hate these things. I told them that the fireplace must be checked. It is true actually. I told them that they were so late that everything was getting damaged. The chimney cap should be constructed as soon as possible.

Tectonic Affects Caused by Changes in the Production and Placement of the Fireplace and Chimney

Two firms and a carpenter were involved in the positioning of the fireplace (Table 6.4). There was a professional firm producing and installing metal fireplaces. There was another professional firm producing details and special paints. There was also a carpenter who made the timber belt for the fireplace.

---

39 Heidegger, *Being and Time*.
40 Hurol, *Tectonic Affects in Contemporary Architecture*.
41 Ingold, *Making – Anthropology, Archeology, Art and Architecture*.

Table 6.4 Tectonic affects due to changes during the placement of the fireplace and construction of the chimney.

The change in 14	The symbol	The tectonic affect(s)	Responsible person/ contributor	Related changes	Innovative attitudes
Having a modern fireplace.		Tectonic affect of time	The RONAC	0	None
Change of the position of the fireplace.		It was badly affecting the space quality and blocking the continuity between indoor and outdoor spaces.	Responsible: the application project. Solution: the RONAC.	Many	None
Changing the dimensions of the fireplace.		Better tectonic detail.	Suggested by the fireplace firm.	0	None
Having a two-directional fireplace.		Poetic tectonic affect as a detail.	Suggested by the RONAC and accepted by the contractor during the contract phase.	0	No, it was an industrialized fireplace.
Having a timber belt around the fireplace and using a special paint on the fireplace.		Haptic tectonic affect due to the use of timber and the color of copper.	Suggested by an architect friend of the RONAC.	Many	None
Elimination of the low chimney height.		Elimination of wrong detail and making chimney visible.	Problem by: the contractor. Solution: the RONAC.	0	None
Design of the chimney cap.		Tectonic detailing in harmony with the house.	Problem and solution by a builder and the RONAC.	0	Semi-realized possibility with tectonic affects.

Eliminated changes/problems are shaded in light gray in the table. Changes which cause another change or originate from another change are shaded in dark gray. Changes that did not cause any other changes are unshaded. Symbols reflect not only the type of change, but also the associated feelings. Lighter colors in symbols reflect positive feelings.

216 | *6 Tectonic Affects and Innovative Attitudes due to Construction Changes*

Tectonic affects due to changes in the installation of the fireplace in the house were usually positive. All of the trouble caused by the application project and the contractor was eliminated by the RONAC, and some other contributing changes were also realized during this process. There were seven changes causing tectonic affects during this process.

The two-sided fireplace facing was made out of black basalt marble. A fourth builder came to do this. "While taking the measurements he threw his cigarette into the fireplace, and I got angry with him."

Reflection(s) from Theory

Tectonic affects of time and change are technical or technological improvements, innovations, having an affirmative approach to technical rules, being new and familiar, being practical, timeliness, an ontological approach to time, the rejection of history, being conservative, and being futurist.[42] The Monarga House combines the features of **being new** and **familiar** in its different parts. The fireplace is one part which reflects the new. Because of the combination of new and old, the Monarga House also has an **ontological approach to time**.[43]

Innovative Attitudes During Construction of the Fireplace and Chimney

- The change of chimney cap triggered an innovative attitude, and this resulted in a unique design (the witch's hat) which required some research. However, it was canceled due to the danger of lightning, and an experienced builder and foreman designed and built a typical but safe chimney cap.
- It is known that birds get into chimneys. The contractor had an innovative attitude about this problem and suggested placing a bird cage on the top of the chimney.

Conclusion

There were many changes during this phase of construction, including those with tectonic affects and innovative attitudes. There were three eliminated changes and many singular changes. The majority of these had positive affects. The changes that were caused by previous changes were all initiated by the application project and tendering process phases. The changes causing tectonic affects are seen in Figure 6.5.

---

42 Hurol, *Tectonic Affects in Contemporary Architecture*.
43 Gevork Hartoonian, *Ontology of Construction – On Nihilism of Technology and Theories of Modern Architecture*. (Cambridge University Press, 1994).

Conclusion | 217

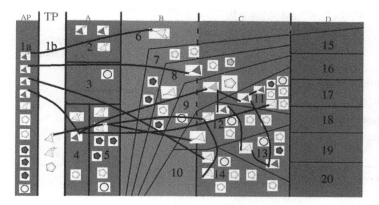

Figure 6.5 The map of the tectonic process for Phase C - Application project (AP, 1a) and tendering process (TP, 1b), construction of foundations (2), the frame system (3), walls and openings (4), the roof (5), plastering and painting (6), the electrical system (7), the mechanical system (8), ceramics (9), water isolation (10), the windows, shutters, and front door (11), the interior doors (12), the wardrobes and cupboards (13), the fireplace and chimney (14), the garden walls (15), drainage (16), the garage and concrete work in the garden (17), metalwork (18), the pathways (19), and timber work in the garden (20).

Table 6.5 The changes during the Phase C of the Monarga House.

Phase C	Eliminated changes	Singular changes	Multiple changes	Total number of tectonic changes	All changes	Innovative attitudes
Windows, shutters, front door (C 11)	1 positive	3 positive, 1 negative	1 positive, 2 negative	5 positive + 3 negative: 8	12	2 realized with tectonic affects, 2 unrealized with tectonic affects, 1 without tectonic affects,
Interior doors (C 12)	1 positive	3 positive	2 positive, 1 negative	6 positive + 1 negative: 7	7	2 realized and both caused tectonic affects,
Wardrobes/ cupboards (C 13)	1 positive	1 positive, 1 negative	1 positive, 1 negative	3 positive + 2 negative: 5	8	1 realized and caused tectonic affects,
Fireplace/ chimney (C 14)	1 positive	4 positive	2 positive	7 positive: 7	9	1 realized, 1 semi-realized, both with tectonic affects,
Total	4 positive	11 positive, 2 negative	5 positive, 4 negative	21 positive + 6 negative: 27	36	6 realized, 2 unrealized, 1 semi-realized (all with tectonic affects), 1 realized without tectonic affects

Table 6.6 The actors initiating changes with tectonic affects in Phase C.

Phase C	Contractor/builder/ foremen/workers	The RONAC/ the main controller
Windows/shutters/front door (11)	1 positive, 2 negative	5 positive
Interior doors (12)	2 positive, 1 negative	4 positive
Wardrobes/cupboards (13)	2 positive, 1 negative	1 positive
Fireplace/chimney	2 positive	5 positive
Total	7 positive, 4 negative: 11	15 positive

Some decisions were made collectively by all parties.

Table 6.5 presents the changes that caused tectonic affects, the total number of changes (including those without tectonic affects), and the number of innovative attitudes during Phase C. The increase in positive tectonic affects may be attributed to the nature of this phase, which can be categorized as a finishing phase where it is easier to implement changes.

The actors who initiated the changes with tectonic affects within Phase C of the construction of the Monarga House are listed in Table 6.6. The contractor, builder, foremen, and workers made 11 changes with tectonic affects, while the RONAC and the main controller made 15 changes.

The reflections from theory within this chapter include tectonics and construction theories, architectural theories, various philosophies, and theories of aesthetics. The dominating theory among these is the theory of tectonics.

The reflections of theory in respect of the changes, tectonic affects, and innovative attitudes during this phase of the building process relate to many different types of theories. These are:

- **Theories of tectonics** such as continuity, tectonic details, natural building materials, hiding the logic of construction, ontology, hapticity, poetic tectonic affects, atectonics, hidden error, tectonics and stereotomics, hearth, tectonic affects of change or time, the new and familiar, and an ontological approach to time
- **Theories of architecture** such as passive house design principles, having dots and lines – as well as elements and components – related to each other, monotonous and meaningless environments, minor architecture, eclecticism, crowded tastes, camouflage, repetition, hierarchy, unity, rhythm, and ambiguity

Conclusion | **219**

- **Various philosophies** such as an instrumental approach, enframing, care, authenticity, a temporary nature, space as clearing out, fetishism, functional and aesthetic detailing, power – hegemony and ignorance – female empowerment, psychoanalysis of fire, and commodity value
- **Theories of construction** such as high-performance windows and doors, change management, compensation for mistakes, change orders, corruption – deception and theft – design during construction, copper – cuprum, cyprium, Cyprus
- **Theories of aesthetics** such as the symbol of the witch
- **Theories of anthropology** such as being close and being distant

7

Tectonic Affects and Innovative Attitudes due to Changes During the Construction Work in the Garden

Many people might think that work on the garden is not a part of the building construction. However, there can be a lot of construction work involved in gardens.

The RONAC was looking for a large garden because she has many pets. She also preferred to have a green garden. When she saw the present land which is approximately 900 m², she decided to buy it because of these reasons. One of the first ideas was to have a house hidden at the back of the mimosa trees in the front part of the garden. This decision eliminates cutting these trees which become totally yellow during spring times.

In order not to lose the sea and mountain view, the RONAC (Researcher, Owner, Neighbor, Architect, Controller) decided to use metal fences on three sides of the garden, while having a solid wall facing the street. The garage was designed as a separate building to create a courtyard in the garden. Three dog kennels were placed in the front part of the garden to ensure better security. Many of the elements in this phase of construction changed a lot during the building process.

Reflection(s) from Theory

The fifteenth of the 17 SDGs (Sustainable Development Goals) demands sustainable cities and communities which require consideration about the **need to include plants, insects, and animals**.[1] The Monarga House hosts many dogs, cats, birds, and a black snake. Sometimes there are foxes and hedgehogs around. The house has a natural garden containing pepper, mimosa, olive and mulberry trees, and many cactuses.

(Continued)

---

1 United Nations, *An Architecture Guide to the UN 17 Sustainable Development Goals*, accessed December 18, 2022. https://www.uia-architectes.org/wp-content/uploads/2022/03/sdg_commission_un17_guidebook.pdf.

Tectonics as a Process in Architecture, First Edition. Yonca Hurol.
© 2025 John Wiley & Sons, Inc. Published 2025 by John Wiley & Sons, Inc.

(Continued)

Mike Davis explained the reasons for the increase in the need for **security precautions** and gave examples of these, including special architectural elements and the image of the building as well as the cameras, security guards, and so on. Davis also wrote that Frank Gehry said that architecture is sufficient to solve security problems and there is no need for cameras, and so on.[2] Many of the security precautions in the Monarga House concern earthquakes, wildfires, flooding, and so on. There are also precautions against uninvited human beings; however, these precautions are not about the image of the building. The image is modest.

Garden Walls – Changes, Tectonic Affects, and Innovative Attitudes

The front garden wall was designed as a white-washed brick or reinforced concrete wall with red tiles on top. However, the contractor preferred to not to use concrete because of the plastering and painting, and so on, and asked the RONAC whether this wall could be built of stone which might not have plaster and paint. The RONAC liked this idea and accepted. As the soil type is rocky, no foundations were built under these stone walls. However, when they started cracking, they had to be repaired and tie-beams were built behind them later.

Since the foundation trench for the building was not dug correctly as previously mentioned, and the RONAC did not want balustrades on the terraces, it also became necessary to create a new level at the back of the building. This level was also surrounded by stone retaining walls approximately 70 cm high. The steps to the terraces were also built in stone. There is another stone wall which divides the garden in two. Figure 7.1 shows the front garden stone wall, other stone walls

Figure 7.1 (a) The front garden wall. (b) Stone wall and steps inside the garden. (c) The supporting concrete tie-beam added to the back of the stone garden wall at the front.

---

2 Mike Davis, *City of Quartz: Excavating the Future in Los Angeles*. (Verso, 2006[1990]).

Garden Walls – Changes, Tectonic Affects, and Innovative Attitudes | **223**

at the back with steps around the terraces and the additional tie-beam behind the front garden stone wall. This idea of the contractor led to the presence of stone in the house.

> When I visited the place, I checked the stone walls. They were lovely, but there were some holes in them. I will ask about these.

Reflection(s) from Theory

Gaston Bachelard wrote about the **tectonics of timber roofs and stone cellars** in his *Poetics of Space* and explained timber and stone spaces. Heavy, cold stone and cellars which are below the earth level, represent death and painful feelings such as sadness and fear in humans.[3]

Stone is only used in the garden of the Monarga House. The stone walls in the garden and the timber used inside and outside of the house, contribute to the haptic character of the house. Juhani Pallasmaa used the concept of **hapticity** especially for natural materials.[4]

Changes During the Construction of Garden Walls[5]

- The use of stone for the front wall, the wall splitting the garden in two, the retaining walls for the new level created around the terraces and the garden steps added a second natural and haptic material to the house.
- The front garden wall did not have foundations. Since the soil is rocky, this decision was made together with the building team (#39). "I am worried about the placement of this wall. I think it should not be on the pavement's tie-beam at all. I remember the main controller warned us about this. I have to call the contractor about this." Through time, and especially after boring a hole under this wall for house's electricity cable, these stone walls started cracking dangerously. The RONAC expressed her feelings about this problem in her diary as follows: "The stone wall at the entrance of my house – the one close to the electric post – is not safe anymore. It has some serious cracks. I think this is especially due to the hole under it. I am worried that this might cause some problems." Later, these walls were maintained and reinforced concrete tie-beams were added at the rear.

---

3 Gaston Bachelard, *The Poetics of Space*, Trans. M. Jolas. (Beacon Press, 1994[1958]).
4 Juhani Pallasmaa, *The Eyes of the Skin – Architecture and the Senses.* (John Wiley, 2005).
5 Hashtags (#) indicate mistakes which could have been identified and solved by experienced foremen.

> *Reflection(s) from Theory*
>
> **The angle and thickness of cracks** are important in deciding about the dangers they may cause for structures.[6] As the RONAC taught structure courses, including the structures of buildings, she understood the dangers stone walls posed in the garden of the Monarga House.

- Since the garden had not been excavated as had been specified in the preliminary project, a level problem occurred on the back terrace of the house. As mentioned earlier, the east and south terraces were too high. "Together with other layers above this level the total height of the brick walls will be approximately 47 cm above the level of the earth at the highest point. This also changed the number of steps required at the back of the house. In the project specification the back terrace had three steps to the garden. But now it will be around six or seven steps. And the direction of these steps should be parallel to the terrace." The contractor suggested adding a balustrade to these terraces. The RONAC did not accept this. The second solution was raising the level of the whole garden. However, this would cause a problem with the tie-beams of the fences which had already been built. Therefore, a stone wall was built to surround the problem terraces and to increase the level of that area only. This took up too much space on the east side of the garden. However, it contributed to the tectonic presence of stone in the house.

> *Reflection(s) from Theory*
>
> **Ensuring that the construction** follows the design project is an important part of control during the construction of buildings.[7] This was not done properly especially at the beginning of the construction of the Monarga House.

- The placement of the stone wall which divided the garden in two was unintentionally slanting and it had to be changed because it was 2 m (#40) into the dog kennel area.

---

6 Petros Christou and Miltiades Elliotis "Construction and Retrofit Methods of Stone Masonry Structures in Cyprus" *The Open Construction and Building Technology Journal* 10, no. Suppl. 2. M6, 2016: 246–258.

7 Saleh Mubarak, *Construction Project Scheduling and Control*. 2nd Edition. (John Wiley and Sons, 2010).

> *Reflection(s) from Theory*
>
> This book mentions many types of errors in architecture such as errors in articulation,[8] benign errors,[9] imperfection of fragile architecture (related to "weak thought"),[10] and errors causing haptic affects.[11] Some of them are consciously done and some are not. Although some of them can be seen as negative by professionals, they all result in good things for humans. However, having a slightly slanting wall might not contribute to the architecture of any building or the lives of any users. Such errors can be regarded as **disturbing errors**.

- The small decorative wall in front of the garage which also holds the letter box, was built too low. But its height was increased after the RONAC's warning (#41).
- Most of the steps outside the house were also made of stone. This also contributed to the tectonic presence of stone in the house (#42).
- The steps on the west terrace were built like a small amphitheater. This was not specified in the original project and it is strange (#43). Who will sit there to watch something?
- When the construction was complete, the RONAC found that the lack of small ramps connecting both sides of the garden caused trouble for her. Eventually, she designed a small portable timber ramp which could be kept in the garage.

> I am a bit worried about the situation of water meter because it was too close to the wall. It might be difficult to make a box for it.

Tectonic Affects due to Changes During the Construction of Garden Walls (Table 7.1)

There were six changes causing tectonic affects while creating the garden walls. Most tectonic affects due to changes in the garden walls originated from incorrect excavation at the beginning of the construction. Therefore, it became necessary

---

8 David Sylvester, *Interviews with Francis Bacon*. (Thames & Hudson, 2016).

9 Pallasmaa, *The Eyes of the Skin*; Alvar Aalto, "Inhimillinen Virhe (The Human Error)." In *Nain Puhui Alvar Aalto (Thus Spoke Alvar Aalto)*, Ed: G. Schildt. (Otava, 1997): 282. Alvar Aalto and Göran Schildt, "Speech at the Centennial Celebration of the Faculty of Architecture" on May 12, 1972 at Helsinki University of Technology." In *Alvar Aalto in his Own Words*. (Rizzoli, 1998).

10 Gevork Hartoonian, *Ontology of Construction – On Nihilism of Technology and Theories of Modern Architecture*. (Cambridge University Press, 1994).

11 Marcel Proust, *In Search of Lost Time (À la Recherche du Temps Perdu)*. 7 volumes. (Everyman's Library, 2001[1913–1927]). The titles of the seven volumes are *Swann's Way*, *In the Shadow of Young Girls in Flower*, *The Guermantes Way*, *Sodom and Gomorrah*, *The Prisoner*, *The Fugitive*, and *Time Regained*.

226 | 7 Tectonic Affects and Innovative Attitudes due to Changes

Table 7.1 Tectonic affects due to changes during the construction of garden walls.

The change in 15	The symbol	The tectonic affect(s)	Responsible person/ contributor	Related changes	Innovative attitudes
Having the garden walls in stone.		Use of a natural material, causing haptic tectonic affects.	Contractor suggested, the RONAC accepted.	0	None
Having additional stone walls around the back terraces because of an undesired excavation.		Use of a natural material, hapticity, achieving tectonic continuity in the application of stone.	Contractor suggested, the RONAC accepted.	Many	This is a matter of design.
Having most garden steps built in stone.		Use of a natural material, hapticity, achieving continuity in the application of stone.	Contractor suggested, the RONAC accepted.	Many	None
Having additional reinforced concrete tie-beams behind the front stone walls for safety. Rework		Since this should have been done in a better way, it causes negative tectonic affects because it looks heavy and unreasonable.	Caused because of a wrong decision of the building team. Decision made by the main controller, builder, and the RONAC.	0	Realized and caused tectonic affects.
Having no level difference on two sides of the stone wall which divides the garden in two.		This can cause negative tectonic affects because it is not reasonable.	Caused by the contractor and due to unwanted excavation.	Many	None
Having the stair on the west terrace like a small amphitheater.		This causes negative tectonic affects because it is not reasonable.	Contractor	0	None

Table 7.1 (Continued)

The change in 15	The symbol	The tectonic affect(s)	Responsible person/ contributor	Related changes	Innovative attitudes
Elimination of the height problem of the decorative wall in front of the garage door. Rework		If this problem was not eliminated, the wall was meaningless.	The RONAC	0	None

Eliminated changes/problems are shaded with light gray in the table. Changes which cause another change or originate from another change are shaded with darker gray. Changes that did not cause any other changes are unshaded. Symbols reflect not only the type of change, but also the associated feelings. Lighter colors in symbols reflect positive feelings.

to have some additional walls and a false wall. However, the use of stone for these walls was the preference of the contractor which was also accepted by the RONAC.

> *Reflection(s) from Theory*
> Neil Leach wrote about the concept of **camouflage**.[12] The false wall in the garden of the Monarga House was not designed as camouflage. However, because of the excavation error at the beginning of the construction, it turned out to be a false wall which serves as camouflage.

Innovative Attitudes During the Construction of Garden Walls

- Adding reinforced concrete tie-beams to the rear of the cracked stone garden walls resulted from an innovative attitude. This decision was made as a result of a conversation between the main controller and an experienced builder. Although the problem was solved, this application caused negative tectonic affects.

    I also had the chance to see the details of the work done by the builder and his team. He told me that he is making stronger tie-beams than the main controller and I suggested. It is almost like a foundation to stop the stone wall from falling into the garden. He was using L-shaped reinforcement for this. He was right.

---

12 Neil Leach, *Camouflage*. (MIT Press, 2006).

228 | 7 *Tectonic Affects and Innovative Attitudes due to Changes*

> *Reflection(s) from Theory*
> The belief in not expecting innovative attitudes from builders and foremen can prevent such ideas from being reflected in their work. Although it is also thought to be good to open up the innovation process to all team members,[13] it is clear that not much is expected from builders and foremen. However, contrary to the above sources, the garden work of the Monarga House contained some **innovative attitudes from the contractor, builders, and foremen**. They include innovations such as retrofitting the damaged stone wall.

Drainage – Changes, Tectonic Affects, and Innovative Attitudes

The site level is below pavement level. The slope is towards the south-east. There is a possibility that other houses may be built on the three pieces of land on three sides of the site and walls may also be built around them. This might result in having walls on all four sides of the garden. Since the RONAC wanted to place the house at the back and in the lower part of the site, there was the possibility of flooding and drainage issues had to be taken seriously. There were various concerns regarding it. The first precaution is to have a drainage well at the lowest point of the garden. The second precaution[14] is to alter the slope in every part of the garden so that the water runs into the drainage well. The third precaution is to provide drainage pipes for areas that trap water. This happened in two places in the garden; on one side of the passage towards the house entrance and in front of the garage. The slope in these water traps should also be arranged in such a way that water will flow to the drainage pipes to move to the other sides of the garden which slope towards the drainage well. At this point, it is also useful to know that stone walls allow water to flow through because of the holes in them. It was also necessary to make holes for water to pass through the reinforced concrete tie-beams under the fences which surround the garden.

Figure 7.2a shows one of the RONAC's sketches relating to the drainage issues on the site. Most of the decisions in this sketch were applied. Figure 7.2b shows the passage towards the entrance with formwork and reinforcement only. Figure 7.2c depicts the same place with drainage pipes which let the water flooding on the left

---

13 Henri Simula and Tuomas Ahola "A Network Perspective on Idea and Innovation Crowdsourcing in Industrial Firms" *Industrial Marketing Management* 43, no. 3, 2014: 400–408. Xiaolong Xue et al., "Collaborative Innovation in Construction Project: A Social Network Perspective" *KSCE Journal of Civil Engineering* 22, 2018: 417–427.
14 United Nations, *An Architecture Guide to the UN 17 Sustainable Development Goals*.

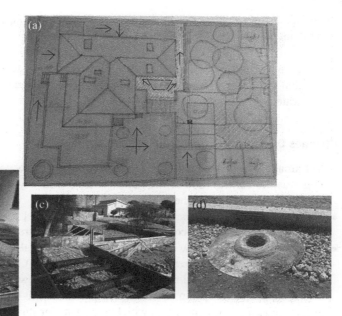

Figure 7.2 (a) The RONAC's sketch regarding decisions related to drainage. (b) Reinforcement and formwork for the concrete parts in the garden. (c) Drainage pipes added below these concrete surfaces in two places in order to let the trapped water flow to the lower parts of the garden. (d) The well in the lowest part of the garden to collect drainage water.

side of the photo flow to the right side which is lower. After this photo was taken, concrete was poured over them and the footpath was made. There is a similar drainage pipe under the garage. Figure 7.2d demonstrates the 10 m-deep drainage well at the south-east corner of the site. Water can be seen when one looks into it. However, it still has not been used as a water well.

> *Reflection(s) from Theory*
> Achieving **sustainability in the use of water and rainwater** is considered in the sixth SDG of the United Nations Guide for Sustainable Goals. It is useful to collect and re-use dirty water as well as to collect and re-use rainwater. Rainwater accumulates in a well in the garden of the Monarga House. However, the black and gray water from the Monarga House is not currently used.

The preliminary project only contained the drainage well which was the contribution of another contractor and some circular pipes to let the water move below

7 Tectonic Affects and Innovative Attitudes due to Changes

the footpath towards the entrance. However, the drainage pipe under the garage was not included in the project.

> I am worried about drainage issues because on the lower side of the garage and at the end of the stone wall there should be another conduit. Otherwise, water will accumulate around the garage.

Changes During the Organization of the Drainage[5]

- A drainage pipe was added under the garage to let the water flow through it. This was the contribution of the main controller. The type of pipe to provide drainage was selected by the contractor.
- The organization of slopes and the placement of gravel where the water was expected to flow were the contribution of the main controller (#43).
- The filter at the top of the drainage well was very weak, but it was changed by the builder after a warning from the RONAC (#44). "I saw that the filter for the drainage well was in place as I requested. I was worried and I also sent its photographs to the main controller, and he said it was correctly done having the two sides of the well prepared with big and smaller stones and covered safely with a fixed metal top. I just want to add some gravel around the well to make it look better."

> *Reflection(s) from Theory*
> **Water wells in both parts of Cyprus** were used to dump the bodies of murdered people during the wars. The well in the Monarga House is closed with a fixed metal top, so that no one can fall into it or open it.

- Later, the main controller found a hole in the opposite direction under the garage and it was filled in with concrete. The main controller said that if such holes are developed further, they can cause cracks in the foundations of the garage.
- Two ends of the drainage pipes were closed with wire mesh to prevent small animals from going into them.
- Later, the RONAC found that in some places around the house the water draining off the roof was hitting the earth and splashing onto the white walls, making them dirty. The same problem was occurring on the floor of the garage. Since the bottom of the garage is open, the splashing water makes the floor of the garage dirty. The RONAC put gravel in these places and solved the problem.

The organization of the drainage was not solved at once. There were several ideas and decisions before reaching the final solution. These also contained the

Drainage – Changes, Tectonic Affects, and Innovative Attitudes | 231

possibility of having two manholes beside the entrance passage and beside the garage. However, the team later decided that this was not necessary, and they were canceled. The rain chains beside the passage to the entrance and on the back terrace also contribute to the water drainage system.

> *Reflection(s) from Theory*
> Martin Heidegger used the concept of **"clearing out"** in order to define what is space.[15] What should be cleared out is an important discussion in architecture. However, it is certain that water should be kept away from buildings. Removing water from the Monarga House was a serious issue due to the characteristics of the site.

Tectonic Affects due to Changes During the Installation of Drainage

Tectonic affects are usually not expected due to drainage. However, the two changes causing tectonic affects within the process of drainage of the Monarga House can be seen in Table 7.2.

Table 7.2 Tectonic affects due to changes during the application of drainage.

The change in 16	The symbol	The tectonic affect(s)	Responsible person/contributor	Related changes	Innovative attitude
Having visible drainage pipes under the garage and the footpath.		Tectonic detail	The RONAC, the main controller, contractor	0	Realized and caused tectonic affects.
Elimination of muddy water splashing on the white walls by putting gravel beside the plinth protection.		Tectonic detail	The RONAC	0	None

Eliminated changes or problems are shaded with light gray in the table. Changes which cause another change or originate from another change are shaded with darker gray. Changes that did not cause any other changes are unshaded. Symbols reflect not only the type of change, but also the associated feelings. Lighter colors in symbols reflect positive feelings.

---

15 Martin Heidegger, "Art and Space" Trans. Charles Seibert, 1969, accessed August, 2023. https://pdflibrary.files.wordpress.com/2008/02/art-and-space.pdf.

Innovative Attitudes During the Installation of Drainage

- Having a drainage pipe under the garage and the type of pipe used for this purpose originated in an innovative attitude of the contractor. However, these decisions were made by all parties involved in the construction.

The Garage and Concrete Work in the Garden – Changes, Tectonic Affects, and Innovative Attitudes

Although they did not exist in the project, the main controller insisted on having plinth protection surface around the building. The RONAC accepted them as long as they were not wider than 60 cm. Since the foundations have already been built, they had to be attached to the building by injecting reinforcement into the side surfaces of the foundations with the help of epoxy. Figures 7.3a–c demonstrate various stages in the construction of these plinth protection surfaces.

The garage had a reinforced concrete frame and brick walls in the preliminary project which changed several times during the building process. The contractor agreed to have a simple metal garage. However, someone had forgotten to write this into the contract. Later, the team decided on a galvanized iron garage. Finally the iron garage was built with narrow openings at the top and bottom of its walls to avoid extreme heat during hot summers.

> *Reflection(s) from Theory*
> Tectonics of the Monarga House can be defined with Gotfried Semper's concept of **stereotomics** and the tectonics of the garage in the same house can be defined with his concept of **tectonics**.[16] This is because the garage is made of iron and is light, but the house has a reinforced concrete frame and brick walls which puts it into the category of stereotomics. There are multiple tectonics in the Monarga House. This concept was introduced by Gevork Hartoonian as a **montage** of different characteristics, elements, and components.[17]

---

16 Gottfried Semper, *The Four Elements of Architecture and Other Writings*, Trans. Harry F. Mallgrave and Wolfgang Herrmann. 2nd Edition. (Cambridge University Press, 2010[1851]). Gottfried Semper, *Style in the Technical and Tectonic Arts; or, Practical Aesthetics*, Trans. Harry F. Mallgrave. (Getty Research Institute, 2004[1860]). Gottfried Semper, *The Four Elements of Architecture and Other Writings*, Trans. Harry F. Mallgrave and Wolfgang Herrmann. 2nd Edition. (Cambridge University Press, 2010[1851]).
17 Hartoonian. *Ontology of Construction*.

The Garage and Concrete Work in the Garden – Changes, Tectonic Affects, and Innovative Attitudes | **233**

Figure 7.3 (a) Steel injection used to build a plinth protection around the house. (b) Construction of steel reinforcement and formwork for the plinth protection. (c). Removal of the formwork after the curing of the plinth protection. (d) Positioning of concrete for the garage and footpath. (e) Construction of the steel frame for the garage. (f) The steel frame covered with a corrugated metal sheet. (g) Position of the aluminum door of the garage. (h) The drainage under the garage and the gap under the pedestrian passage. Steel was injected and concrete was poured into this gap to solve the problem. (i) Flowers painted on the garage.

The foundation slab of the garage and the reinforced concrete footpath to the entrance had many construction problems. They were not at right angles with the building, and because of this, the ceramics looked very strange on them. The ceramics highlighted the angle differences. Finally, an additional triangular concrete surface was installed by injecting reinforcement into concrete plus epoxy and the problem was solved. Later, the garage was painted by an artist so that

7 *Tectonic Affects and Innovative Attitudes due to Changes*

it did not resemble military barracks. Figures 7.3d–i show various stages of the construction of these concrete surfaces and the metal garage.

Reflection(s) from Theory

Certainty is expected in modern buildings because they should be constructed in accordance with professional projects. However, **extreme precision** can cause many problems. An example of this is the house designed by the philosopher Ludwig Wittgenstein for his sister. He based everything in the house on mathematical proportions and ignored the functional and ergonomic requirements of its users. The door handles are not in a suitable place, as they are positioned in the middle of the door. It is also known that Wittgenstein asked workers to pull down a reinforced concrete slab because it was a few millimeters lower than it should have been. This building did not have the character of a home. There were many problems like this, because Wittgenstein tried to achieve absolute mathematical perfection.[18] However, rejection of the slanting footpath and the arrangement of its ceramics in the Monarga House do not originate from an extremely perfectionist idea.

Having the garage painted by an artist camouflages it.[19] However, this does not make the garage look like a hobbit house. It is just some flowers painted on the garage.

Changes During the Construction of the Plinth Protection[5]

- Since both the preliminary and application projects did not include plinth protection, this issue is totally new. The RONAC did not want to have too much concrete around the house. However, the main controller warned the RONAC that the site was rocky, which would cause water problems and soil the house walls. Both the RONAC and the contractor accepted this, and a narrow plinth protection was built around the house. However, there were some ugly construction mistakes concerning the especially at the back of the house (#45). Since this plinth protection is narrow, the freely falling rainwater still caused earth to splash onto the white walls for a while. To solve this problem, gravel

---

18 Stuart Jeffries, "A Dwelling for the Gods" *The Guardian*, January 5, 2002, accessed February 22, 2025. https://www.theguardian.com/books/2002/jan/05/arts.highereducation.
19 Leach, *Camouflage*.

The Garage and Concrete Work in the Garden – Changes, Tectonic Affects, and Innovative Attitudes | **235**

was placed beside these surfaces. In a way, this change originated from the change in roof type from which water falls freely.

> *Reflection(s) from Theory*
> The fifteenth Sustainable Development Goal relates to **protecting nature** and taking precautions for this purpose.[20] The Monarga House has a large and almost natural garden (by having naturally growing trees and plants) and minimizes the use of concrete especially in the garden.

- A detail problem between the plinth protection and the terrace steps was spotted and solved by a worker. That is his contribution (#46).

  We thought that it would be possible to have a reinforced concrete slab for the foundations of the garage. However, since these foundations will be placed on the filled-in area, I am now worried about it. I called the main controller but he did not answer his phone.

Changes During the Construction of the Garage and Work with Concrete to Provide Pedestrian Footpaths[5]

- There was an angle problem relating to the footpath connecting the garage to the house (#47). Because of this, when the ceramics were placed at the correct angles, a gap occurred under them. This was solved by injecting steel into concrete and pouring concrete into that gap.
- Since the house was designed to be accessible, the garage must be 50 cm higher to be level with the entrance of the house. There should not be any steps between the garage and the house. Since the excavation was done incorrectly at the beginning, the front part of the site needed some filling to achieve accessibility. This required the foundations of the garage to be higher. The team thought that it would be possible to have a reinforced concrete slab as the foundations of the garage. The slab for the garage foundations had to be deeper than expected and they had more reinforcement.

---

20 United Nations, *An Architecture Guide to the UN 17 Sustainable Development Goals.*

236 | *7 Tectonic Affects and Innovative Attitudes due to Changes*

> *Reflection(s) from Theory*
>
> The tenth of the 17 SDGs, demands the reduction of inequalities by considering human rights, including the rights of disabled people. This can be reflected in architecture by providing **accessibility** for disabled people and universal design to consider the needs for everybody.[21] The accessible character of the Monarga House was protected during the construction of the buildings and walls in the garden.

- As mentioned previously, the garage was changed several times during the design and construction processes. There were also attempts to change the location of the garage. Some people thought it should not be at the back of the plot. Other people thought that it should not be blocking the sea view. Its location has not been changed. It does not block the sea view, but blocks the view from neighboring houses which is good. However, it is necessary to open two doors to enter and leave the garage.
- The main controller slightly increased the dimensions of the eaves of the garage roof at the design stage to stop water getting into the garage.
- The builders arranged the position of the aluminum door of the garage by placing the iron frame elements accordingly. This was the idea of one of the workers. However, the door placed by the aluminum firm was slanting (#48).

> *Reflection(s) from Theory*
>
> This book mentions errors in architecture such as errors to articulate,[22] benign errors,[23] imperfection of fragile architecture (weak thought),[24] and errors causing haptic affects.[25] However, errors such as having the slanting door might not improve any quality for humans. Instead, they are **disturbing errors.**

- The selected green-colored corrugated sheet could not be found for the garage. The firm suggested white. However, the RONAC did not accept this. The choice

---

21 United Nations, *An Architecture Guide to the UN 17 Sustainable Development Goals.*

22 Sylvester, *Interviews with Francis Bacon.*

23 Pallasmaa, *The Eyes of the Skin*; Aalto, "Inhimillinen Virhe (The Human Error)." Aalto and Schildt, "Speech at the Centennial Celebration of the Faculty of Architecture."

24 Hartoonian, *Ontology of Construction.*

25 Proust, *In Search of Lost Time (À la Recherche du Temps Perdu).*

of a shade of green for the garage instead of white meant that it could be hidden in nature. Another problem, concerning the military green color chosen, was that it was not available in the same form as the other colors. Its sheets had to be cut by hand rather than by machine. When cut by machine, the cut edges are automatically painted thus reducing any rusting problems later. However, when cut by hand this is not possible and it can cause rusting. Later, one of the RONAC's architect friends told her that there will not be any serious rusting problem because water cannot stay on these surfaces.

- Flowers were painted on the garage because of its color and size.
- Concrete was added to the floor at the front of the garage entrance and to the sides of the metal garden gates. This was necessary because the roads and paths in the garden are of stabilized earth.

Reflection(s) from Theory
The fifteenth Sustainable Development Goal relates to **protecting nature** and taking precautions for this purpose.[26] Minimizing concrete in the garden of the Monarga House achieves this target to a certain extent.

Tectonic Affects due to Changes During the Construction of the Garage and Concrete Work in the Garden

The six changes causing tectonic affects during this phase of construction can be seen in Table 7.3. Decisions about the garage were changed several times. The material of its structure and roof changed three times. Its color was changed twice. Its location was questioned, but not changed. Finally, the garage was completed without harming the original design decisions. The plint protection around the building did not exist in the preliminary and application projects.

A year later "I started to feel uncomfortable about the general image of the garage. It is smaller than the kennels, but it looks like military barracks. Especially from far away."

Innovative Attitudes During the Construction of the Garage and the Concrete Work in the Garden

- The decision to add reinforced concrete plint protection to the building was a result of an innovative attitude of the main controller.

---

26 United Nations, *An Architecture Guide to the UN 17 Sustainable Development Goals.*

238 | *7 Tectonic Affects and Innovative Attitudes due to Changes*

Table 7.3 Tectonic affects due to changes during the construction of the garage and concrete work in the garden.

The change in 17	The symbol	The tectonic affect(s)	Responsible person/ contributor	Related changes	Innovative attitudes
Having reinforced concrete plint protection with some bad details around the building.		Against the concept of the house, atectonic detail. However, it eliminates water problems and dirt – due to water coming from roof – on the façade.	Responsible people: Contractor, workers	Many	Realized and caused tectonic affects
Solution to having an angle problem with the footpath to the entrance which also caused bad placement of ceramics. Rework		This could have caused negative tectonic affects because the detailing was poor and unreasonable.	Responsible people: builders, workers, the RONAC Solution: the builder, the main controller	0	Same solution with the previous one
Having an iron garage with openings at the top and bottom.		Parallel to the concept of minor architecture, which allows other creatures to enter.	The building team	0	None
Having the iron garage in green.		Parallel to the concept of being hidden in nature.	The RONAC	0	None
Solution to having an slanting aluminum garage door for the garage.		A timber element was added to hide this bad detail due to the inclination of the door.	Responsible people: the builder Solution: the RONAC	0	None
Solution to the military appearance of the garage because of its color and size.		Attempts were made to eliminate this problem by painting flowers to make it more human.	Responsible people: the contractor, but the RONAC also accepted	0	None

Eliminated changes and problems are shaded in light gray in the table. Changes which cause another change or originate from another change are shaded in darker gray. Changes that did not cause any other changes are unshaded. Symbols reflect not only the type of change, but also the associated feelings. Lighter colors in symbols reflect positive feelings.

Fences, Dog Kennels, Garden Gates – Changes, Tectonic Affects, and Innovative Attitudes

These changes can be referred to as changes in metalwork. The type of fence changed several times during the construction. However, finally it was built like the fence in the preliminary project. There are reinforced concrete stepped tie-beams underneath the fences, and these are factory-produced metal fences. This type of fence was the most appropriate to be able to appreciate the sea and the mountain views. The garden gates were designed as factory-produced metal gates in the preliminary project. However, they were built as specially produced metal gates.

> *Reflection(s) from Theory*
>
> There is a difference between the **tectonics of industrialized products and specially produced products** (like traditional products and buildings) because there is usually no strict precision in the specially produced products. This makes them look more natural. An example can be given by comparing modern timber balloon frames with traditional timber framed masonry. Traditional timber framed masonry buildings contain varying element dimensions and varying distances between them. The Monarga House contains both industrialized and specially produced elements and components.

When the house was designed the RONAC only had two dogs. However, the number of dogs increased to four. Since one of the dogs is free to roam in the garden, three kennels were built. Therefore, many changes also occurred in the construction of these elements of the project.

Figures 7.4a and b demonstrate the formwork of the stepped tie-beams of fences and the fences after their completion. Figures 7.4c and d show one of the dog kennels and the metalwork on the windows of the front door. Later, the dog kennel roofs were covered with a hard surface.

Changes due to Metalwork in the Garden[5]

- The fence type was changed several times during the construction. The design project included factory-produced fences. Later, the use of a certain type of mesh was preferred. There were also many different ideas about how they could be placed. Finally, they were built as factory-produced fences.

Figure 7.4 (a) The formwork in place for the stepped tie-beams for the metal fences. (b) The metal fences in place. (c) The metal dog kennels. (d) The metalwork placed against the glass panel beside the front door for security.

- The original project did not contain tie-beams for the fences. Building tie-beams was suggested by the contractor during the tendering process while preparing the contract agreement. This is a firm way of building fences. Fence tie-beams are stepped at the front of the site, but they are not at the back of the site. This has caused sloped fences at the back. "The contractor put steps (for the tie-beams of fences) to the front part of the garden, but he left them slanting in the rest of the garden and that the fences are also slanting. This looks strange."

> *Reflection(s) from Theory*
> Tectonics theories were written to explain the **tectonic characteristics of visible parts of buildings**. Similarly, contractors, builders, foremen, and workers give more importance to the parts at the front and visible elements of buildings.

Fences, Dog Kennels, Garden Gates – Changes, Tectonic Affects, and Innovative Attitudes | **241**

- The metal dog kennels were built too high (#49). "I am worried because they are too high, higher than the garage and the front part of the site looks military. I am not sure if I really suggested this height … I should have asked for the same height as the garage or something lower than the garage." Since the RONAC was abroad and she realized this after the kennels were built, it was not possible to change them. The dog kennels being higher than the garage made a negative impact on the relationship between the masses on the site. This also made the kennels very visible from the street. The front façade of the site became like a prison. The RONAC managed to grow attractive trees close to the stone walls and kennels within three years and these trees concealed the height of the kennels to a certain extent.

> *Reflection(s) from Theory*
>
> **Proportions of building parts and their relationship** with each other (addressing each other) are very important in architecture.[27] The kennels being higher than the garage at the Monarga House disturbed the proportions of the buildings in the project.

- In the original project the kennel floors were earth. However, considering the possibility of the dogs digging to escape and the smell caused by earth floors, the RONAC decided to have the floors of the dog kennels concreted. However, this increased the amount of concrete on the site.

> *Reflection(s) from Theory*
>
> The fifteenth Sustainable Development Goal relates to **protecting nature** and taking precautions for this purpose.[28] Although the RONAC wished to have a natural garden with minimum concrete cover, the floors of the dog kennels were concreted to prevent the dogs from escaping and to minimize the smell.

- The problem with the location of dog kennels and the garage door was spotted and solved by a worker (#50).
- The metalwork on the windows of the main front door was designed by the RONAC during construction. This became necessary because the front door was not produced as expected. The builder brought silver and gold paint and the RONAC asked for them to be mixed to paint the curved elements in the metalwork.
- It was decided to have handmade metal garden gates instead of factory-made ones as specified in the preliminary project.

---

27 Francis D.K. Ching, *Architecture, Form and Space*. (John Wiley and Sons, 1996[1979]).
28 United Nations, *An Architecture Guide to the UN 17 Sustainable Development Goals*.

Table 7.4 Tectonic affects due to changes during the construction and placement of metalwork.

The change in 18	The symbol	The tectonic affect(s)	Responsible person/contributor	Related changes	Innovative attitudes
Having a more professional and transparent fence with stepped tie-beams.		Looks professional and enables a view of the sea. Its lightness is tectonically significant.	The RONAC	Many	No, this is typical
Having slanting tie-beams and fences in the back of the garden.		They look as if they were unprofessionally built. Bad detailing.	Contractor	0	None
Having the dog kennels too high.		They are seen from outside and contrast with the garage which is very low. Bad proportions.	Responsible people: the builder	0	None
Having concrete floors for the dog kennels.		Against the concept of the project because it increases the amount of concrete.	Responsible people: the RONAC	0	None
Having ornamental metalwork on the windows of the front door. Rework		Tectonic detail	Responsible people: the RONAC	Many	Realized and caused tectonic affects.
Having handmade metal garden gates.		Tectonic detail	Responsible people: the main controller, the RONAC	0	None

Eliminated changes/problems are shaded in light gray in the table. Changes which cause another change or originate from another change are shaded in darker gray. Changes that did not cause any other changes are unshaded. Symbols reflect not only the type of change, but also the associated feelings. Lighter colors in symbols reflect positive feelings.

Tectonic Affects due to Changes During the Construction and Placement of Metalwork in the Garden

The contractor had been legally dismissed at this stage and a specialist firm dealt with all types of metalwork and applying it on site. It was a quick process. The six changes causing tectonic affects during this phase of construction can be seen in Table 7.4.

Innovative Attitudes During the Construction and Placement of the Metalwork

- The metalwork on the openings of the front door was designed by the RONAC. However, it was changed by the builder with an innovative attitude to avoid some problems. Although this metalwork was needed for security, it was not possible to attach it to the marble to prevent it from cracking. Then, the safety of the metalwork was achieved by the strength of its bars.

Pathways and Roads – Changes, Tectonic Affects, and Innovative Attitudes

The two vehicle roads (the garage entrance and septic tank entrance) on the site were planned as stabilized earth. Stone pieces had to be placed on both sides to keep the stabilized earth in place. The pedestrian footpath was also designed with stabilized gravel to minimize the amount of concrete on the site. After moving into the building, the RONAC added some more pathways by using standard pieces of concrete. Figures 7.5a–c demonstrate the stabilized earth vehicle roads on the site and earth and gravel in the garden. Figures 7.5d–f show some pathways in the garden. Figures 7.5g and h show the laundry area in the garden and the changes made to the pavements beside the road.

> When I saw them under the west August sun in the afternoon at around 3 o'clock without hats, I was really worried. The face of one of the boys was red.

Changes in the Pathways and Roads[5]

- The contractor warned the RONAC that the house did not have any place for a septic pumper truck to reach the septic tanks at the back at the beginning of

244 | 7 Tectonic Affects and Innovative Attitudes due to Changes

Figure 7.5 (a) One of the stabilized earth vehicle entrances. (b). Stone pieces placed against the two sides of the stabilized earth roads. (c) Earth and gravel placed in the garden. (d–f) Pathways in the garden. (g) The laundry area. (h) The badly rebuild pavement in front of the house to enable vehicles to get in and out.

the construction. Therefore, a second vehicle entrance was added. This entrance was also built with stabilized earth just like the main vehicle entrance.
- The materials of the stabilized garden roads were changed. The project specified a yellower earth road. However, Havara soil was also acceptable. The organization of two types of stabilized earth (gravel and Havara soil) in the garden was the contribution of the builder.

- The footpath in the garden was designed as a free curve inspired by the curved pathway between the former pigeon and chicken coops. However, it was drawn as a large rigid path in the application project. Finally, it was realized as a free curve as it is in the preliminary project.

Reflection(s) from Theory

Aldo Rossi believed that architecture should belong to its context and every context has its own history. Therefore, architecture should fulfill its place by **relating to the history of its context**. He researched the history of the sites of his buildings and designed them with an awareness and respect for that history. He also rebuilt the lost architecture as his architecture.[29] The site of the Bonnafantenmuseum was an industrial area parallel to the Maas River. The area also traverses the old city center. Rossi carried out research on these old industrial buildings, which no longer existed, and designed his project to reflect their characteristics. The museum building is symmetrical. There is a monumental dome, which is accessed by a monumental staircase located in the middle of the building. The roof over the stairs is glass, and this gives an outdoor feel to the location of the stairs. Rossi used traditional materials (stone, brick, timber) and modern materials (concrete and steel) together in this building.

The Monarga House relates to the context by having similar features to the old buildings in Iskele (Trikomo), such as leaf plaster and green shutters as well as similarities to some old buildings in the same village. The house also relates to the context by having a pathway like the one which was between the former chicken coops on the same land.

- The stabilized ground beside the entrance of the house was realized with gravel and nylon was put under it to avoid plants growing. Gravel was used both for the footpath and the area beside the entrance of the house. This was a good idea from the builder to make these materials address each other (#51). Having two types of earth-based natural materials was also good.
- It was planned to fill in the garden with red earth. However, the builder warned the RONAC that red earth is difficult to manage because it makes everything dirty (#52). So brown earth was used instead.
- Apart from the gravel pathway from the entrance gate to the garden, there were no garden pathways in the original project. However, it was not practical to walk on mud on rainy days. Therefore, prefabricated stone pieces were used to make simple pathways in the garden.

---

29 Aldo Rossi, *The Architecture of the City*. (MIT Press, 1982).

> *Reflection(s) from Theory*
>
> Martin Heidegger suggested that people should leave the highways and main roads of thought and lose themselves in the **pathways** of forests. He said that they should walk until they see a place with light coming through the trees (the light of being). Then they should stop and spend some time there, but not too long ... They should continue walking again.[30]
>
> Heidegger's pathways are pathways of thought. However, they are analogical to natural pathways. The major garden pathways of the Monarga House are earth-based and naturally curved. Secondary pathways were built after moving into the house and spending some time in it. Determining the place and form of secondary pathways once the buildings are in use is a common idea in architecture. This idea adds a more natural feature to the environment. It also creates a comfortable and reasonable environment for the users.

- The west terrace steps were designed like an amphitheater by the contractor much earlier. This was different from the project and it is strange. This place has no similarity to any of the other terraces or steps. It is foreign to the building. However, at least it is at the rear of the building. This happened because it was the first flight of steps to be built for the house.

> *Reflection(s) from Theory*
>
> **Amphitheaters look towards the north** so that the sun doesn't blind the audience. The amphitheater faces west in the Monarga House.

- The pavements beside the street were rebuilt in a bad way (#52). Some cement blocks were broken, and the workers had to cover many places with concrete.
- As some trees had collapsed, they were cut up. This decreased the hidden character of the house for a while.
- A laundry area was formed in the garden with galvanized posts and wires. This is similar to the preliminary project.

Tectonic Affects due to Changes During the Construction of Roads and Pathways in the Garden

Most of this work was done by a specialist builder and again without the presence of any contractor. The five changes causing tectonic affects in this activity can be seen in Table 7.5.

---

30 Martin Heidegger, "Thinker as Poet." In *Poetry, Language, Thought*, Trans. Albert Hofstatter. (Harper and Row, 1975[1947]): 1–14.

Pathways and Roads – Changes, Tectonic Affects, and Innovative Attitudes | **247**

Table 7.5 Tectonic affects due to changes during the construction of roads and pathways in the garden.

The change in 19	The symbol	The tectonic affect(s)	Responsible person/ contributor	Related changes	Innovative attitudes
Having two different types of stabilized earth.		Creating tectonic variety with similar materials.	The builder	0	Realized and caused tectonic affects.
Having a second vehicle entrance.		This was functionally necessary and aesthetically pleasing because it is stabilized earth.	Contractor	0	None
Having a curved and natural footpath in the garden.		Parallel to the idea of having a natural area in the front part of the garden, was tectonic continuity with nature.	The RONAC	0	None
Having additional pathways in the garden of standard concrete blocks.		Creating tectonic variety with a similar concept to other roads and pathways.	The RONAC	0	None
Having the steps of the west terrace like an amphitheater.		Against the project and not reasonable. Bad detail.	The first contractor	0	None
Rebuilding the municipal pavement badly.		Bad tectonic detail, careless work.	The builder, workers	0	None
Canceling the use of red earth.		Better to have cleaner surfaces.	The builder	0	None

Eliminated changes/problems are shaded in light gray in the table. Changes which cause another change or originate from another change are shaded in darker gray. Changes that did not cause any other changes are unshaded. Symbols reflect not only the type of change, but also the associated feelings. Lighter colors in symbols reflect positive feelings.

I am worried that they might harm the underground water pipes while working with their bulldozers.

Innovative Attitudes During the Construction of the Roads and Pathways

- The change in the material of the stabilized earth and the use of two different materials for stabilization was due to an innovative attitude of the builder.

Timber Work in the Garden – Changes, Tectonic Affects, and Innovative Attitudes

The original project specified timber only on the terraces. The pergola posts, pergolas, and fascia on the roof eaves were designed as being made of timber. However, after the first contractor agreed to compensate for the heat insulation on the reinforced concrete frame with timber interior door frames and after replacing the aluminum sliding door between the entrance hall and the living area with a timber sliding door, the RONAC started to think that there could be more timber in and around the house. Especially after terminating with the contractor, the RONAC used every opportunity to use timber, including in the garden. Having a good new carpenter also contributed to this situation. This carpenter built one part of the two rain chains, shields for the dog kennels, the water tank and the metal gates. These shields were needed to decrease contact with the neighbors. However, the maintenance of timber surfaces is a problem on the coast. So, it is no longer a good excuse to exaggerate the amount of timber work. Figure 7.6 shows the shields around the water tank, metal gates, and the dog kennels.

> Today the carpenter came and completed the garage gate. He placed the timber pieces vertically and he used shiny silver buttons to cover the screws. I love it.

Figure 7.6 (a) The timber screen around the water tank. (b). Timber screen on the metal garden gates. (c) Timber screen for the dog kennels.

Changes During the Execution of Timber Work in the Garden

- Three types of timber screens in the garden were not planned in the original project. These screens made the house even more isolated from the street and caused haptic tectonic affects due to the use of natural materials. These are the screens for the water tank, dog kennels, and garden gates. These relate to the timber work in the house including the interior door frames, the timber belt around the fireplace, the timber trap door to the attic and timber covering material on the pergolas.

> *Reflection(s) from Theory*
> As mentioned early, Gaston Bachelard wrote about the **tectonics of timber roofs and stone cellars** in his "Poetics of Space" and explained that timber which is an easily workable material, and the attic which is nearest the sky, represent the human mind.[31] The use of timber in architecture also causes **haptic tectonic affects**.[32]
> The use of timber in the Monarga House was important for the house.

- The timber screen on the water tank protects the tank from the effects of the strong western sun.

Tectonic Affects due to Changes During the Construction of Timber Elements in the Garden

The three changes during this process, which caused tectonic affects, can be seen in Table 7.6.

> *Reflection(s) from Theory*
> As mentioned earlier, Neil Leach wrote about the concept of camouflage and said that **camouflage** in architecture indicates a desire to be connected.[33] The shields and screens of the Monarga House were positioned to provide more protection for the animals and to separate them from the neighbors who were disturbed by their presence. In a way, they were placed there to achieve better relationships with these neighbors.

I was actually worried about having too much timber.

---

31 Bachelard, *The Poetics of Space*.
32 Yonca Hurol, *Tectonic Affects in Contemporary Architecture*. (Cambridge Scholars Publishing, 2022).
33 Leach, *Camouflage*.

250 | *7 Tectonic Affects and Innovative Attitudes due to Changes*

Table 7.6 Tectonic affects due to changes during the construction of timber elements in the garden.

The change in 20	The symbol	The tectonic affect(s)	Responsible person/ contributor	Related changes	Innovative attitudes
Having a timber screen around the water tank.		Use of a natural material, hapticity, achieving continuity in the application of timber.	The RONAC	1	None
Having a timber screen on one side of the dog kennels.		Use of a natural material, hapticity, achieving continuity in the application of timber.	The RONAC	1	None
Having a timber screen on the metal garden gates.		Use of a natural material, hapticity, achieving continuity in the application of timber.	The RONAC	1	None

Eliminated changes/problems are shaded in light gray in the table. Changes which cause another change or originate from another change are shaded in darker gray. Changes that did not cause any other changes are unshaded. Symbols reflect not only the type of change, but also the associated feelings. Lighter colors in symbols reflect positive feelings.

Innovative Attitudes During the Execution of the Timber Work in the Garden

- Although there was no innovative attitude while the timber work in the garden was being done, the design of the timber pieces of the rain chains results from an innovative attitude of the carpenter as well as one of the builders and the RONAC. However, this was listed in Chapter 4 as an innovative attitude under the title Roof and Pergolas.

Conclusion

There were many changes during this phase of construction, including those with tectonic affects and innovative attitudes. There were four eliminated changes and many singular changes. The majority of these had positive affects. The changes caused by previous changes were all initiated by the application project, tendering process, and the changes made during the construction of the foundations. All of these changes, which caused tectonic affects, are shown in Figure 7.7. This figure represents the entire map of tectonic affects for the Monarga House.

Conclusion

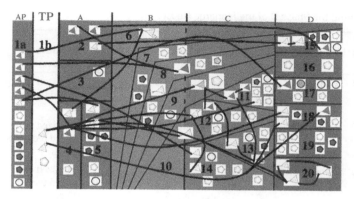

Figure 7.7 The map of the tectonic process for Phase D - Application project (AP, 1a) and tendering process (TP, 1b), construction of foundations (2), the frame system (3), walls and openings (4), the roof (5), plastering and painting (6), the electrical system (7), the mechanical system (8), ceramics (9), water isolation (10), the windows, shutters, and front door (11), the interior doors (12), the wardrobes and cupboards (13), the fireplace and chimney (14), the garden walls (15), drainage (16), the garage and concrete work in the garden (17), metalwork (18), the pathways (19), and timber work in the garden (20).

Table 7.7 presents the changes causing tectonic affects, the total number of changes (including those without tectonic affects), and the number of innovative attitudes during Phase D. The increase in the number of positive tectonic affects might be due to the nature of these items. They are in the garden, making it easier to change them. Tectonic affects caused by the architectural elements in the garden complete the overall tectonic affect of the building. Since these are the last elements of construction it is more probable that changes in them to originate from previous changes. The stone walls and timber elements in the garden have been considerably affected by the previous changes in the construction. Although they did not exist in the project, they have made a considerable contribution to the tectonic qualities of the building. Most of these items were built after the RONAC and the contractor terminated their agreement and because of this their production was better controlled than the earlier phases.

In contrast to the high number of changes with tectonic affects, the innovative attitudes are fewer than in the previous phases. These innovative attitudes will be useful in delaying the deterioration of the house.

The actors who initiated the changes with tectonic affects during Phase D of the Monarga House construction are listed in Table 7.8. The contractor, builder, foremen, and workers made 19 changes with tectonic affects, while the RONAC and the main controller made 17 changes.

The reflections from theory in this chapter include theories of tectonics and construction, architectural theories, theories about sustainability, various philosophies, theories about structural engineering, and politics. The dominant theory among these is, once again, the theory of tectonics.

252 | *7 Tectonic Affects and Innovative Attitudes due to Changes*

Table 7.7 The changes during the Phase D of the Monarga House.

Phase C	Eliminated changes	Singular changes	Multiple changes	Total number of tectonic changes	All changes	Innovative attitudes
Garden walls	1 positive	1 positive, 2 negative	2 positive, 1 negative	4 positive + 3 negative, total: 7	8	1 realized and caused tectonic affects.
Drainage	None	2 positive	None	2 positive	6	1 realized and caused tectonic affects.
Garage and concrete work	2 positive, 1 medium	2 positive	1 negative	4 positive + 1 neutral + 1 negative: 6	10	1 realized and caused tectonic affects.
Metalwork	None	1 positive, 2 negative	2 positive, 1 negative	3 positive + 3 negative: 6	7	1 realized and caused tectonic affects.
Pathways	None	5 positive, 2 negative	None	5 positive + 2 negative: 7	10	1 realized and caused tectonic affects.
Timber elements	None	None	3 positive	3 positive	2	None
Total	3 positive, 1 neutral	9 positive, 6 negative	7 positive, 3 negative	19 positive + 10 negative: 29	43	5 realized and caused tectonic affects.

Table 7.8 The actors initiating changes with tectonic affects in Phase D.

Phase D	Contractor/builder/ foremen/worker	The RONAC/the main controller
Garden walls (15)	3 positive, 3 negative	1 positive, 1 negative
Drainage (16)	2 positive	1 positive
Garage and concrete work (17)	2 positive, 1 neutral, 1 negative	4 positive, 1 medium
Metalwork (18)	2 negative	3 positive, 1 negative
Pathways (19)	3 positive, 2 negative	2 positive
Timber work in the garden (20)	—	3 positive
Total:	10 positive, 1 neutral, 8 negative: 19	14 positive, 1 neutral, 2 negative: 17

Note that some decisions was made collectively by all parties.

Conclusion | **253**

The reflections of theory

- *Theories of tectonics:* such as the tectonics of timber roofs and stone cellars, disturbing errors, stereotomics and tectonics, montage, visible parts of the building, tectonics of industrialized, specially produced products, and certainty and extreme precision
- *Theories of architecture:* security precautions, hapticity, orientation of amphitheaters, camouflage, reflecting on the history of the context and proportions of parts and their relations
- *Theories about sustainability:* such as protecting plants and animals, use of water, protecting nature, and accessibility.
- *Various philosophies:* such as space as clearing out and pathways of thinking.
- *Theories of construction:* such as controlling and innovative attitudes.
- *Theories about structural engineering:* such as the angle and thickness of cracks in stone walls.
- *Politics:* such as the water wells in Cyprus.

Part III

Changes, Tectonic Affects, and Innovative Attitudes within the Building Process

Part III contains two chapters that present all the findings of the book, seeking to achieve conclusions from the specific case of the Monarga House and seven interviews which were conducted with professionals. Chapter 8 finalizes the process maps for the tectonic affects and innovative attitudes and discusses them. These process maps illustrate the interrelationships between processes and the changes that give rise to tectonic affects and innovative attitudes as presented in the previous chapters. In other words, Chapter 8 presents the results of the research in this book and visualizes them through these process maps and accompanying tables. Guided by the results of all the chapters and the insights provided by the maps and interviews, the book culminates in the conclusive Chapter 9. The conclusion explores the potential contributions of this research to tectonics theories by offering an in-depth discussion of the building process of Monarga House. This unique perspective is made possible by the RONAC, who has observed the process from multiple viewpoints: as a Researcher, Owner, Neighbor, Architect and Controller.

Tectonics as a Process in Architecture, First Edition. Yonca Hurol.
© 2025 John Wiley & Sons, Inc. Published 2025 by John Wiley & Sons, Inc.

8

Process Maps for Tectonic Affects and Innovative Attitudes within the Building Process

This chapter discusses the following issues to reflect on the results of the research about the building process of the Monarga House:

- The entire map of changes causing tectonic affects during the building process, considering their components (multiple changes that trigger other changes, eliminated changes as outcomes of good control procedures, and singular changes that may independently occur at any time during the building process)
- The map of innovative attitudes, considering each as a possibility that is either realized or not realized
- Actors of changes that cause tectonic affects (whether contractor, builder, foreman, technician, carpenter, and worker, the RONAC (Researcher, Owner, Neighbor, Architect, Controller) and the main controller, the architect of the application project, or all actors)
- Theoretical reflections in Chapters 2–7 related to changes, changes with tectonic affects, innovative attitudes, and actors of change, as they are placed in relevant parts within these chapters.

The Map of Changes with Tectonic Affects for the Monarga House

The changes within the building process, those causing tectonic affects and those initiating innovative attitudes are presented in Chapters 3–7. Figure 8.1 presents the entire map of changes with tectonic affects based on the analysis presented in these previous chapters. This map combines multiple changes that trigger other changes, eliminated changes that protect the architectural design decisions, and singular changes that can also improve the tectonic quality of the building.

Tectonics as a Process in Architecture, First Edition. Yonca Hurol.
© 2025 John Wiley & Sons, Inc. Published 2025 by John Wiley & Sons, Inc.

8 Process Maps for Tectonic Affects and Innovative Attitudes within the Building Process

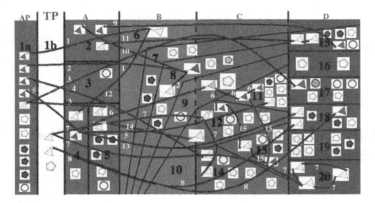

Figure 8.1 The map of the tectonic process of Monarga House with an emphasis on multiple changes. Application project (AP, 1a) and tendering process (TP, 1b), construction of foundations (2), frame system (3), walls and openings (4), roof (5), plastering and painting (6), the electrical system (7), the mechanical system (8), ceramics (9), water isolation (10), the windows, shutters, and front door (11), interior doors (12), wardrobes and cupboards (13), the fireplace and chimney (14), garden walls (15), drainage (16), garage and concrete work in the garden (17), metalwork (18), pathways (19), and timber work in the garden (20).

The multiple changes in this figure are connected to each other with lines and white numbers are assigned to the related sets of these lines. The changes that take place within these sets of multiple changes are as follows:

1) From AP (application project) negative: "The contradiction between the heat insulation on the structural elements and the leaf plaster was left ambiguous." To 4 (walls, openings, and heat insulation) positive: "Having heat insulation on structural elements became impossible, but this enabled the application of leaf plaster to the walls." To 6 (plaster) positive: "Application of leaf plaster on certain walls."

2) From AP (application project) negative: "Change of tank capacity from 4 tons to 2 tons." To 8 (mechanical systems) negative first: "Decisions about the placement and capacity of the main water tank: from 2 tons (negative) to 5 tons later (positive)." To 8 (mechanical systems) positive: "The final installation of a 5 ton water tank in a good location."

3) From AP (application project) positive: "Roof gutters were canceled and the roof was designed for water to fall freely." To 17 (the garage and concrete work in the garden) negative: "Reinforced concrete plinth protection with some poor details were added around the building to avoid water problems caused by freely falling water from the roof" and to 4 (roof, pergolas) positive: "Two rain chains were added to the building to control water coming from two roof hip lines."

The Map of Changes with Tectonic Affects for the Monarga House | 259

4) From AP (application project) negative: "The fireplace was moved forward to open up more space in the dining area." To 14 (fireplace and chimney) positive: "Change of the position of the fireplace" to avoid blocking openings in the living area.

5) From AP (application project) negative: "Accordion windows and shutters were eliminated due to poor construction quality in North Cyprus." To 4 (walls, openings, heat insulation) negative: "The heights of the windows were not built according to the project specification due to changes in the window and shutter types." To 11 (windows, shutters, entrance door) positive: "Semi-transparent openings were added above and to one side of the front door." The change in window type affected the window heights and due to the change in the lintel heights in all openings, it became necessary to add a transom window above the front door.

6) Also related to 5 – from 4 (walls, openings, heat insulation) Negative: "The heights of windows were not done according to the project." To 9 (Ceramics) medium: "Tiles were changed in the kitchen and bathroom to create continuity between the different height levels of doors, windows, and cupboards" also to 11 (windows, shutters, entrance door) negative: "Not having fully openable windows in the living area due to the window and shutter type and the difference in floor levels between inside and outside" and "Incorrect window heights" and to 12 (interior doors) negative: "The height of interior doors being lower than that of the windows."

7) From TP (tendering process) positive: "Change of the timber cover under the eaves of the roof and pergolas." To 5 (roof, pergolas) positive: "Use of timber in many areas. This started with the change of material and post connection details in the pergolas and continued with the interior door frames." Also to 12 (interior doors) positive: "Having the interior door frames made of timber." To 14 (fireplace, chimney) positive: "Having a timber belt around the fireplace and applying special paint on the fireplace." Also to 20 (timber elements in the garden) positive: "Having a timber screen around the water tank" and "Having a timber screen on one side of the dog kennels" and "Having a timber screen on the metal garden gates."

8) From TP (tendering process) positive: "Having stepped reinforced concrete tie-beams under the garden fences." To 18 (metalwork) negative: "Having inclined tie-beams and fences at the back part of the garden" and to 18 (metalwork) positive: "Having a more professional and transparent fence partially with stepped tie-beams."

9) From 2 (foundations) positive: "Having stone walls around the building due to faulty excavation." To 15 (garden walls) negative: "Having no level difference on the two sides of the stone wall that divides the garden in two."

8 *Process Maps for Tectonic Affects and Innovative Attitudes within the Building Process*

10) From 2 (foundations) negative: "Not opening a hole at the tie-beam level to allow a pipe to reach the toilet manhole." To 8 (mechanical systems) negative: "All the manholes, especially the toilet manhole, were higher than expected."

11) From 2 (foundations) positive: "Having stone walls around the building due to faulty excavation." To 15 (garden walls) positive: "Having additional stone walls around the back terraces due to incorrect excavation" and "Having most garden steps built in stone."

12) From 4 (walls, openings, heat Insulation) positive: "Having heat insulation on structural elements became impossible, but this allowed for leaf plaster on the walls." To 6 (plaster, painting) positive: "Application of leaf plaster on certain walls."

13) From 4 (walls, openings, heat insulation) negative: "The heights of the windows were not built according to the project." To 18 (metalwork) positive: "Ornamental metal work was added to the windows of the front door." These windows did not exist in the original project. They appeared because the lintels of all the openings were set at the wrong level.

14) Also related to 7 from 5 (roof, pergolas) positive: "Use of timber in many places related to the roof. This started with the change of material and post connection details in the pergolas and continued with the interior door frames." To 12 (interior doors) positive: "Small transom windows were added over the interior door frames because the timber frames of the interior doors were too high."

15) Also related to 5. From 11 (windows, shutters, entrance door) negative: "Incorrect window heights were determined." To 13 (wardrobes, cupboards) negative: "The height of wardrobes and cupboards was different from that of the windows and doors" and positive: "Solutions developed to the gaps above the bedroom wardrobes" and to 12 (interior doors) negative: "The height of the interior doors was lower than that of the windows." Various solutions were implemented for the gaps above the wardrobes and cupboards: one involved adding a niche, another involved extending the wardrobe doors upward and the final solution was using ceramic tiles to improve their appearance.

Although the RONAC identified 15 related sets of changes within this map, after analyzing them, she decided that three of them could be included in other sets of changes. Therefore, there were 12 related sets of multiple changes in the entire construction process of Monarga House. These multiple changes originated from the following phases of the building process:

- Eight originated from the application project (AP).
- Three originated from the tendering process (TP).
- Three originated from the foundations (2).
- Three originated from the walls, openings, and heat insulation (4).

Table 8.1 Changes causing tectonic affects at different phases of the Monarga House.

	Phase 1	Phase A	Phase B	Phase C	Phase D	Total
With positive affects	7	4	8	21	19	59
With negative affects	7	3	4	6	10	30
With neutral affects	–	–	2	–	–	2
All changes	23	36	36	36	43	174

These results show that ambiguities or changes within these early phases triggered many changes in the later phases of the building process. This highlights the importance of the early phases of building processes within a procedural theory of tectonics, as there is a greater chance of changes and the creation of ambiguous issues during these phases.

Table 8.1 demonstrates the total number of changes and changes with tectonic affects for all phases of Monarga House. The high number of total changes (174) indicates that many changes occur in places where they cannot be perceived after the competition of the building.

There are 30 negative tectonic affects, 59 positive tectonic affects, and 2 neutral tectonic affects, making a total of 91 changes with tectonic affects. Fourteen of the negative tectonic affects occurred before Phase C and 16 of the remaining negative tectonic affects occurred later. Nineteen of the positive tectonic affects occurred before Phase C and 40 of the remaining positive tectonic affects occurred later. This shows that the number of positive tectonic affects increased during the later phases of the building process. One factor determining this could be the nature of these activities, as they are mainly related to details and construction work in the garden.

Figure 8.2 presents the map of only the eliminated changes within the process. It is also noteworthy that some negative changes could not been eliminated. A clear example of this is the incorrect excavation of the site, which triggered many changes listed above as the eleventh set of multiple changes. The majority of eliminated changes are positive because they protect the project and design ideas as results of good control.

The list of eliminated changes are as follows:

- In AP, 1a application project: Connecting the terraces to each other was eliminated. This was contrary to the design project, which featured separate terraces in all directions.
- In 3 reinforced concrete frame system: Change in the column height was eliminated.
- In 5 roof and pergolas: The decrease in the height of the roof due to a change in the angle of the roof elements was eliminated.

8 Process Maps for Tectonic Affects and Innovative Attitudes within the Building Process

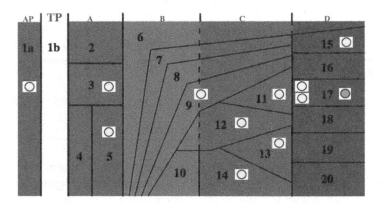

Figure 8.2 Tectonic process map showing only the eliminated changes. Application project (AP, 1a) and tendering process (TP, 1b) construction of foundations (2), the frame system (3), walls and openings and heat insulation (4), roof (5) plastering and painting (6), the electrical system (7), the mechanical system (8), ceramics (9), water isolation (10), windows, shutters, and front door (11), interior doors (12), wardrobes and cupboards (13), fireplace and chimney (14) garden walls (15), drainage (16), garage and concrete work in the garden (17), metalwork (18), pathways (19), and timber work in the garden (20).

- In 9 ceramics: The wall tiles in the kitchen and bathroom were changed twice due to incorrect placement. Their lines did not match and they had to be replaced.
- In 11 windows, shutters, and entrance door: Having three rails for the large aluminum windows instead of two provided more continuity between indoor and outdoor spaces.
- In 12 interior doors: Eliminating the incorrect cornice details with gypsum and the use of silicone on the door windows was necessary. The wrong cornice details could have caused cracks over time and the use of silicon on interior door windows was a poor tectonic detail.
- In 13 wardrobes and cupboards: The elaborate (fancy) kitchen cupboard handles were replaced.
- In 14 fireplace and chimney: Elimination of the low chimney height, which affected the building mass and this was unable to draw smoke properly.
- In 15 garden walls: Elimination of a very low decorative wall in front of the garage door, which made the wall meaningless. The height of the wall was increased.
- In 17 garage and concrete work in the garden: (1) Solution to the angle problem at the pedestrian footpath to the entrance, which also caused improper placement of ceramics. (2) The solution to the inclined aluminum garage door was by covering the problematic part. (3) The solution to the military appearance of the garage, due to its color and size, was to paint some flowers.

The Map of Changes with Tectonic Affects for the Monarga House | **263**

These eliminated changes can be grouped based on their impact on the architecture of the building.

- Those related to the mass and image of the building: Changing the height of the columns in 3, the roof height (which also eliminates the triangular roof windows) in 5 and the height of the chimney in 14.
- Those affecting the spaces in and around the building: Connecting the terraces to each other in 1a, having three rails for aluminum windows instead of two to provide better continuity between indoor and outdoor spaces in 11, increasing the height of the decorative wall in front of the garage to make it more effective in the garden spaces in 15 and addressing the military appearance of the garage due to its color and size by painting flowers in 17.
- Those eliminating bad tectonic details: Changing the wall tiles in the kitchen and bathroom twice due to incorrect details and mismatched lines in 9, eliminating the wrong cornice details with gypsum and the use of silicone on the door windows in 12, replacing the unwanted kitchen cupboard handles in 13, addressing the angle problem of the footpath to the entrance that caused poor ceramic placement in 17 and solving the issue of the slanting garage door by adding a timber piece to the top in 17.

The eliminated changes in the early phases of the building significantly affected its form and image. The eliminated changes in the later phases were mostly about details. It is worth noting that the relocation of the fireplace was decided during the application project, but it was changed again during Phase D, just before its installation. This was a multiple change and it had a significant impact on the interior spaces.

Figure 8.3 presents the singular changes in the process of building the Monarga House. There were many positive and negative singular changes.

The rate of negative tectonic affects within singular changes is higher during the earlier phases of Monarga House, including the application project, tendering process, and Phase A. The rate of positive changes increases in the later phases. This might be due to the nature of those phases, or it might be linked to the termination of the contract between the RONAC and the building contractor. The RONAC believed that the termination of the contract also increased the number of multiple changes in response to earlier changes during the later phases.

The list of singular changes in each phase is as follows:

- In the application project (1a): (1) A small window was added to the west façade. (2) One of the two beams dividing the living area was removed. (3) Different floor levels were used in the terraces and interior spaces to eliminate water penetration (negative). (4) Traditional roof tiles were replaced with modern ones (negative). (5) The placement of AC units was not determined in an architectural manner (negative).

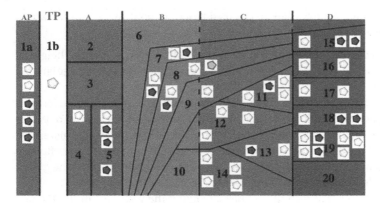

Figure 8.3 Tectonic process map showing only the singular changes. Application project (AP, 1a) and tendering process (TP, 1b), construction of foundations (2), the frame system (3), walls and openings (4), the roof (5), plastering and painting (6), the electrical system (7), the mechanical system (8), ceramics (9), water isolation (10), the windows, shutters, and front door (11), the interior doors (12), the wardrobes and cupboards (13), the fireplace and chimney (14), the garden walls (15), drainage (16), the garage and concrete work in the garden (17), metalwork (18), the pathways (19), and timber work in the garden (20).

- In the tendering process (1b): Change in the connection detail of the timber pergola posts to the floor.
- In walls, openings, and heat insulation (4): The height of windows from the floor was lowered for better ventilation.
- In roof and pergolas (5): (1) Timber posts and other timber elements were installed in pieces (negative). (2) Smaller triangular windows were used (negative). (3) The high ridge tiles were covered with concrete on both sides. (4) The varnish on the posts and fascia on roof eaves was damaged due to delays (negative).
- In electrical systems (7): Poor placement of AC units (negative). (1) Poor placement of the electricity manhole (negative). (2) Addition of a cooker hood to the kitchen. (3) Addition of solar lamps to the garden.
- In mechanical systems (8): (1) Placement of the septic tanks was determined. (2) A 2-ton water tank was suggested (neutral result). (3) The vent pipe for the toilet was added later. (negative) (4) The manhole for the water meter had to be enlarged (negative).
- In ceramics (9): Terraces were edged with marble of a different color.
- In windows, shutters, and entrance door (11): (1) An additional metal element was added between the aluminum windows and the floor tiles. (2) The color of the shutters was changed to a lighter green to reduce the visibility of dust. (3) Two windows overlap with the doors when opened (negative). (4) The color of the windowsills was changed.

The Map of Changes with Tectonic Affects for the Monarga House | **265**

- In interior doors (12): (1) The frames and panels of the interior doors are different from each other. (2) A timber sliding door was installed.
- In wardrobes and cupboards (13): (1) The cupboard for the gas cylinder was forgotten (negative). (2) The kitchen tap was placed in front of the window mullion to prevent it from being visible from the outside.
- In fireplace and chimney (14): (1) Installing a modern fireplace. (2) Changing the dimensions of the fireplace. (3) Installing a two-directional fireplace. (4) Designing the chimney cap.
- In garden walls (15): (1) Constructing the garden walls with stone. (2) Adding reinforced concrete tie-beams behind the front stone walls for safety (negative). (3) Building the stair on the west terrace in the style of a small amphitheater (negative).
- In drainage (16): (1) Visible conduits under the garage and the footpath. (2) Prevent muddy water splashing the white walls by placing gravel beside the plinth protection.
- In garage and concrete work in the garden (17): (1) An iron garage with openings at the top and bottom. (2) The iron garage was painted green.
- In metalwork (18): (1) The dog kennels were built too high (negative). (2) The dog kennels had concrete floors (negative). (3) Handmade metal garden gates were installed.
- In pathways (19): (1) Two different types of stabilized earth were used. (2) A second vehicle entrance was added. (3) A curved, natural footpath was created in the garden. (4) Additional pathways of standard concrete blocks were added in the garden. (5) The steps of the west terrace were built like an amphitheater (negative). (6) The municipal pavement was rebuilt poorly (negative). (7) The use of red earth was canceled.

As seen from the above list, there are some singular changes in materials (e.g., using stone and timber), changes in colors (e.g., shutters, garage, windowsills) and improper placement of some mechanical and electrical systems. There are also some changes related to design decisions, such as having different floor levels inside and outside, using smaller triangular windows, installing a modern two-sided fireplace, designing the chimney cap and building a steps like an amphitheater's. Additionally, there are changes in detailing, such as having different frames and panels of interior doors, altering the joint detail of the timber pergola posts, having these posts in pieces, changing the distance between windows and the floor level, adding a metal element between window rails and floor tiles and having windows and doors overlap when opened, among others.

There are various reworks among these changes. Some reworks are categorized as eliminated changes, such as adjusting the column height by replacing the reinforcement and altering the roof height. Despite the associated loss of time and money, these changes have positive tectonic affects. Reworkings were also required in several stages of the ceramic work and the installation of the aluminum front door. Additionally, there were some reworkings in the Monarga House that did not involve any demolition, such as the addition of a tie-beam to support the stone garden wall. This indicates that not all reworking leads to demolition and their impact on the construction varies.

Impact of Changes on the Initial Design Ideas

Chapter 3 (which presents the design decisions and subsequent characteristics of the Monarga House) and Chapters 3–7 (which present the process of changes and their impact on the tectonic character of the house) demonstrate that the changes in the house's character can be analyzed by comparing the realization of early design decisions about the project with the newly acquired features of the house.

The initial design decisions for Monarga House and the changes to them were as follows:

- Having a living area opening in four directions: Openness to the four directions have **decreased** due to the types of windows and shutters and the difference in floor levels between indoor spaces and terraces.
- *Having a central fireplace*: This decision was **successfully** implemented as designed.
- *Achieving comfort*: The achievement of comfort has **decreased** due to a lack of heat insulation on the reinforced concrete elements and incomplete implementation of the cold roof.
- *Using leaf plaster*: The final result in the use of leaf plaster is an **improvement**.
- *Having an accessible house*: This design objective was **achieved** as intended.
- *Having a small and inexpensive house to run*: The goal of a small and economical house was **partially achieved** according to the original project.
- *Having the house concealed at the back of the garden*: Although the building's height contradicts this design idea, it was **realized** as intended in the design project.
- *Being mindful of the sea view*: The goal of preserving the sea view was according to the initial design.
- *Providing security*: Security measures have **increased** with the addition of cameras and solar lights.
- *Emphasizing the use of various colors*: There has been an **increase** in the use of various colors.
- *Inclusion of tectonic details*: There has been an **increase** in the presence of tectonic details, exemplified by features like rain chains.

Some subconscious dimensions in the design of the house include:

- *Supporting minor architecture*: The inclusion of new materials such as stone and timber, the use of new colors and acceptance of different styles have **increased** the support for minor architecture.
- *Not being domestic*: The subconscious aim of not having a domestic feel has been **achieved**.
- *Presence of old memories*: Preserving and recalling old memories has been **realized**.

Some characteristics that emerged after the initial design of the house are:

- Being open to the contributions of builders/foremen/workers
- Having more topographic levels on the site
- Increase in hapticity due to the use of stone and timber.

Given the substantial changes in the building compared to the initial project, the alterations during the building process have influenced not only the tectonic characteristics but also the design decisions. Although there were many changes with negative tectonic affects, many changes with negative impact were eliminated. The most problematic changes, which were avoided, were either eliminated, or triggered other changes that solved or attempted to solve the problem. Some of the eliminated changes that occurred during the early stages of construction should be highlighted, as they could have seriously affected the form and overall image of the building. The elimination of shorter columns (change in 3) and the lower roof, which would have resulted in the loss of the triangular windows (change in 5) are among these.

Changes that were not eliminated in time triggered additional changes to resolve the issues. Examples include increasing the height of the chimney (change in 14) and changes in topography due to incorrect excavation, which led to the construction of stone walls around the south and east terraces to avoid the need for balustrades (multiple change no. 2 in Figure 8.1). Figure 8.4 compares the east façade of Monarga House as it appeared in the preliminary design project with the potential negative changes in the same façade. These changes could have significantly impacted the form and image of the building. One should also imagine the

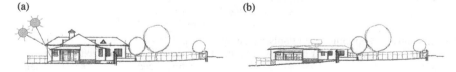

Figure 8.4 (a) The east façade of the building in the preliminary project. (b) One possibility for the east façade that has been eliminated.

left façade with leaf plaster and the right façade with ordinary plaster, along with a structure on the roof to carry a water tank and a boiler, as there is no longer space for them in the attic. Even sunlight could be blocked during the winter if the roof and pergolas were lower.

These changes which could have seriously impacted the planned tectonic character of the house, demonstrate the importance of control, especially in the early phases of construction and the need for changes in the later phases to resolve problems that have already arisen. It is also important to note that many changes in details have contributed significantly to the reality of the building.

Changes with Innovative Attitudes

There were many innovative attitudes (changes through technical knowledge and experience) during the construction of the Monarga House. Table 8.2 presents the innovative attitudes across different phases of the house. Notably, the quantity of all innovative attitudes and those with tectonic affects are nearly the same for Phases C and D. One reason for this could be the nature of these phases, which primarily involve finishing work or construction tasks in the garden.

It should also be taken into account that many changes and innovative approaches do not cause any tectonic affects. Note that Table 8.1 shows 91 changes with tectonic affects, while the total number of changes is 174. Similarly, the number of innovative attitudes causing tectonic affects is 27, while the total number of innovative attitudes is 42, according to Table 8.2. For example, there were many changes during the roof and pergola construction. However, if you check the roof phase in the process map shown in Figure 8.1, you will find that it caused only 7 changes with tectonic affects. This is due to the nature of the roof construction, where many details are hidden and not visible after the

Table 8.2 Changes with innovative attitudes in different phases of the Monarga House.

	Phase 1	Phase A	Phase B	Phase C	Phase D	Total
All innovative attitudes	8 but only 5 realized	11 but only 7 realized	8 but only 5 realized, 1 partially realized	10 but 7 realized, 2 unrealized, 1 partially realized	5, all realized	42, but 29 realized, 2 partially realized
Innovative attitudes with tectonic affects	5 but only 2 realized	5 but only 1 realized	3 but only 1 realized	9 but 6 realized, 2 unrealized, 1 partially realized	5, all realized	27, but 10 realized, 1 partially realized

building's completion. Therefore, many of the changes within this phase do not result in tectonic affects. However, these changes can impact the overall quality of the building and future deterioration of building parts may cause negative tectonic affects over time. Changes and innovative attitudes that occur during construction may also affect other aspects of the building, such as functionality, comfort, safety, and durability, in addition to tectonics.

Figure 8.5 shows the map of innovative attitudes overlaid on the map of tectonic affects. This figure is created by placing signs of realized and unrealized innovative attitudes onto the markers of tectonic affects, where they are related to those changes. If they are not related to tectonic affects, the signs representing innovative attitudes were placed in open spaces on the map. This map depicts the partially realized innovative attitudes as fully realized to avoid further complexity. Although complex, it provides insight into the intensity of these attitudes. Nearly half of the innovative attitudes (19 of 43) stem from multiple changes that also cause tectonic affects, represented by items connected by lines. Ten of them (10 of 43) arise from changes that do not result in tectonic affects. Only two (2 of 43) are linked to eliminated changes, while 12 (12 of 43) come from singular changes that cause tectonic affects. This means that 33 of the innovative attitudes are related to changes with tectonic affects, some of which are positive and others negative.

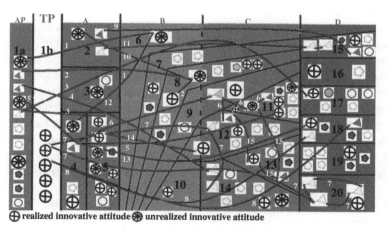

Figure 8.5 Innovative attitudes map drawn on the tectonic affects map. Application project (AP, 1a) and tendering process (TP, 1b) construction of foundations (2), the frame system (3), the walls and openings (4), the roof (5) plastering and painting (6), the electrical system (7), the mechanical system (8), ceramics (9), water isolation (10), the windows, shutters, and front door (11), the interior doors (12), the wardrobes and cupboards (13) the fireplace and chimney (14), the garden walls (15), drainage (16), the garage and concrete work in the garden (17), metalwork (18), pathways (19), and timber work in the garden (20).

Therefore, multiple changes with tectonic affects triggered almost half of the innovative attitudes in the case of the Monarga House.

Preparation of similar maps during construction at various stages can help about predicting changes that may trigger other potential changes in the future. This could lead to discussions about alterations in the building's image, design attitudes, and cost. Although it is not the intention of this book, familiarity with the concept of process maps for changes and tectonic affects may inspire construction firms to develop new change management strategies, aided by computer programs such as BIM (Building Information Modeling).

Actors of Changes with Tectonic Affects During the Building Process

There were many builders, foremen, and workers employed by the contractor during the first part of the construction of Monarga House and many others working with the RONAC later. Table 8.3 presents the actors who initiated the changes causing tectonic affects during the construction of the house. This table was prepared based on 20 tables (a table for each of the 20 stages) presented in Chapters 3–7, each of which lists the actors responsible for the changes with tectonic affects, one by one.

It was more effective to categorize the contractors, builders, foremen and workers together on one side, and the RONAC and the main controller on the other, because the roles of the people on the contractor's side changed throughout the process. The changes initiated by the contractor, builders, foremen, and workers account for (37 of 83) 45% of the total changes with tectonic affects, making them the most influential actors in the changes to the building's tectonic characteristics. However, more than half of the changes they made resulted in negative tectonic affects. The changes caused by the main controller and the RONAC account for (33 of 83) 40% of the total changes with tectonic affects, the majority of which were positive. Only (9 of 83) 11% of all changes were determined by all actors collectively.

These changes, which occurred during the construction of Monarga House and the tectonic affects they caused, serve as an example that supports the claim of this book: the need to understand tectonics and tectonic affects as products of building processes that significantly impact the tectonics of the preliminary architectural project. There can be a discussion regarding the number of such changes. It is reasonable to believe that the number of changes to the Monarga House is too high. However, also supported by the seven interviews, it is known that changes are possible and quite common in all projects.

It is known that various techniques are used by controllers to document and manage changes during construction activities. Sometimes, it is sufficient to have

Actors of Changes with Tectonic Affects During the Building Process | 271

Table 8.3 Actors who initiated the changes causing tectonic affects in the construction process of Monarga House.

	Contractor/ builder/ worker	The main controller and the RONAC	Application project	All actors	Total
Phase A	3 positive, 6 negative	3 positive	None	2 positive	8 positive, 6 negative
Phase B	2 positive, 1 neutral, 2 negative	4 positive, 1 neutral	1 positive, 2 negative	1 positive	8 positive, 2 neutral, 4 negative
Phase C	7 positive, 3 negative	12 positive	1 negative	2 positive	21 positive, 4 negative
Phase D	7 positive, 6 negative	12 positive, 1 neutral	None	3 positive, 1 negative	22 positive, 1 neutral, 7 negative
Total	19 positive, 1 neutral, 17 negative	31 positive, 2 neutral:	1 positive, 3 negative:	8 positive, 1 negative:	59 positive, 3 neutral, 21 negative:
	37	33	4	9	83

reports for each phase of construction. However, at other times, controllers may require documentation of changes, their dates, possible effects, actors involved, and so on. Such documentation might also lead to change orders, which can affect the application project, the cost of the building, and the contract, with the approval of all parties.[1] Although there were significant change attempts during the construction of the Monarga House, such change management activities were not applied.

The 52 changes that could have been or were made by builders, foremen, technicians, and carpenters (not by contractors and subcontractors) to the Monarga House are marked with hashtags in Chapters 4–7. These changes primarily address construction errors (unintentional mistakes) or improvements. They do not require consultation with civil, mechanical, or electrical engineers, nor do they alter the architectural concept of the Monarga House. Additionally, they do not affect the contractor or builder in terms of time or cost. Of these changes, 17 were made by various site actors, 33 had positive or negative tectonic affects and 9 reflected innovative approaches. This demonstrates that allowing builders, foremen, technicians, and carpenters to make decisions about certain types of changes (ones that do not affect engineering systems and building code

---

1 Qi Hao et al., "Change Management in Construction Projects" *International Conference on Information Technology in Construction* CIB W78 2008. (Santiago, Chile, 2008).

requirements, do not alter the architectural concept of the building and do not affect the contractor or builder in terms of cost and time) can be reasonable and beneficial. The above mentioned 52 changes with hashtags were within this category.

Theoretical Reflections for the Changes, their Tectonic Affects, and Related Innovative Attitudes

This book contains many theoretical reflections. Table 8.4 demonstrates the origin of these reflections (such as theories of tectonics, architecture, philosophy (particularly aesthetics, ethics, and politics), construction, anthropology, sustainability, literature, structural engineering) with respect to different phases of construction.

Table 8.4 shows that not only construction, structural engineering, and tectonics, but also other disciplines (including architecture, sustainability, philosophy, and anthropology) relate to almost all phases of the building process of the Monarga House. The presence of theoretical reflections on architecture, philosophy, sustainability, and other areas also indicates that the changes during the process are not only technical but also tectonic.

The Sustainable Development Goals (SDGs) exemplified in this book include Goal 16 (focusing on avoiding economic crisis and poverty), Goal 13 (addressing

Table 8.4 Origin of theoretical reflections with respect to different phases of the building process.

Phases	Origins of theoretical reflections
Context (Chapter 2)	Philosophy (particularly politics and aesthetics), sustainability, construction, architecture, structural engineering
Pre-design	Tectonics, architecture, philosophy, sustainability
AP	Architecture, construction
TP (Chapter 3)	Construction
Phase A (Chapter 4)	Philosophy (particularly ethics and aesthetics), tectonics, architecture, construction
Phase B (Chapter 5)	Philosophy, construction, tectonics, ethics, architecture
Phase C (Chapter 6)	Tectonics, architecture, philosophy (particularly aesthetics), construction, anthropology
Phase D (Chapter 7)	Tectonics, architecture, sustainability, philosophy (particularly politics), construction, structural engineering

The order of theories reflects the number of theoretical reflections.

The Results of the Seven Interviews | **273**

climate action and energy considerations), Goal 11 (targeting the elimination of vegetation loss), Goal 15 (concerning the inclusion of plants, animals, and insects), Goal 6 (related to water usage) and Goal 10 (emphasizing accessibility). Additionally, the book discusses Goal 9 (promoting sustainable building technologies), Goal 1(focusing on avoiding economic crisis), and Goal 12 (advocating for responsible production and recycling).

The Results of the Seven Interviews

The interviews with the OAC (Owner, Architect, Controller), ONACC (Owner, Neighbor, Architect, Controller, Contractor), and OCCC (Owner, Civil engineer, Controller, Contractor), all of whom were closely involved the production process of one building, support the hypothesis presented in this book that "tectonics is/can be a process." They highlighted numerous changes that occurred throughout the course of their projects. The results of these interviews are presented in Table 8.5.

Since the above information is based on the memory of OAC, ONACC, and OCCC, it is expected that there are fewer changes compared to the Monarga House, which was observed and recorded in real-time by the RONAC. Both Case 1 and Case 3 involved changes that impacted the tectonics and the architecture of the buildings. The ONACC commented, "... it is unusual to see changes in the architecture of a building during construction, but this occurred with my house. On the other hand, it is normal to have tectonic changes, such as alterations in materials/systems/details, during the construction process. This is beneficial because on-site conditions and opportunities should be leveraged...." In contrast, Case 2 experienced changes affecting only the tectonics, primarily in the details. This could be attributed to the OCCC's experience as a civil engineering controller, as well as the fact that he served as the contractor of his own house.

Interviews with four other professionals (including an experienced contractor, two experienced controllers, and the head of the Chamber of Architects in North Cyprus) also supported the hypothesis presented in this research. They generally accepted that changes within the building process are normal and beneficial. However, all agreed that these changes should be carried out under the supervision of professionals. OCCC highlighted that changes to the structural system must be overseen by structural and civil engineers. For example, adding a long-span slab into a gallery space, requires engineering knowledge and design. The fourth interviewee, who is an architect, mentioned a case where the builder changed the position of a balustrade over an eave, causing significant harm to the building's architectural image without permission. The architect insisted on reworking the alteration, but the contractor argued that "the detail in the

Table 8.5 Results of interviews with OAC, ONACC, and OCCC.

Case	Eliminated changes	Multiple changes	Changes with tectonic affects	Changes with innovative attitudes	Actors of changes
Case 1 – OAC (2021–2023) ***44 changes**; • 32 with tectonic affects, • 12 without tectonic affects.	3; • eave over the entrance, • sloping walls, • drainage problem and ceramics.	5 series; • 2 about roof, • 8 about the location of the building, • 3 about heating system, • 2 about drainage and ceramics, • 2 about eave over entrance.	Total: 32 • 16 positive, • 13 negative, • 3 medium.	5; • eave over entrance, • the heating system, • the water tank, • about drainage and ceramics, • the front door.	• 17 by the OAC and the main controller, • 15 by the contractor/builders/–workers, • 3 by the civil engineer, • 2 by a technician, • 2 by AP architect, • 1 by City Planning Office.
Case 2 – OCCC (2004–2009) ***13 changes**; • 8 with tectonic affects, • 5 without tectonic affects.	1; • the mistake about re- formwork.	9 series; • 3 at foundation, structural system, and drainage, • 2 at fireplace and ceramics stages, • 2 at electrical and mechanical systems in kitchen, • 2 during the garden work.	Total: 8 All positive.	6; • about earlier phases of construction, • electrical systems, • doors/windows, cupboards in kitchen.	• 8 by the OCCC, • 2 by an artist, • 3 by the OCCC and the electrical engineer.
Case 3 – ONACC (2011–2017) ***16 changes**; • 11 with tectonic affects, • 5 without tectonic affects.	1; • about the low-quality front door.	4 series; • about brutal concrete walls, • collection of roof water, • the water well, • the front door.	Total: 11 • 5 positive, • 5 negative, • 3 medium.	7; • about brutal concrete walls, • addition of fireplace, • water collection, • front door.	3 by the contractor – (contractor was active only at the beginning) • 8 by ONACC, • 1 by a foreman, • 2 by a sub-contractor firm, • 1 by ONACC, electrical engineer, and a technician.

Table 8.6 Results of interviews with a contractor, a controller, and the head of Chamber of Architects in North Cyprus.

Contractor	Number of changes	Multiple changes	Changes with tectonic affects	Changes with innovative attitudes	Actors of changes	Special notes
C1: Architect for 24 years, Contractor for 18 years, designed/ built in North Cyprus, Turkey, England, Dubai, and North Iraq.	17; • in 3 different cases that they were involved in, • in all the buildings designed and built by their company.	8; • e.g., a change in the height of the floor system occurred due to the switch to a floor heating system.	9; • 3 of these involve a new project for an existing structural system, • changing the façade design during construction, adding a steel extension to the top.	7; • cutting a floor in the hall and adding a new one due to a headroom problem.	Foremen can make more changes in small projects. In larger projects, they must follow the hierarchy to implement changes.	They have not observed many changes in the projects they designed and built.
C2 (OCCC): Civil engineer for 40 years, experienced controller, contractor of 3 or 4 buildings.	16; • 13 from Case 2 • 3 more from other buildings.	2; • 1 change in the formwork to eliminate a mistake, • 1 change in a large building to close a gallery with a slab.	10; • 8 from Case 2, • 1 change in the formwork to eliminate a mistake, • 1 to close a gallery with a slab.	6; • 3 from Case 2, • all of the 3 additional changes.	• 11 by the OCCC, • 2 by an artist, • 3 by the OCCC and an electrical engineer (13 comes from Case 2)	He believes that it is natural to have changes during construction, and they are beneficial.

(Continued)

Table 8.6 (Continued)

Contractor	Number of changes	Multiple changes	Changes with tectonic affects	Changes with innovative attitudes	Actors of changes	Special notes
C3: Architect for 15 years, experienced controller of many small and large buildings.	7 changes in different cases.	2; • 1 change because of putting balustrade to the end of the eave, • 1 change due to need for retrofitting.	6; • changing material of a staircase, • 2 about changing place of balustrades, • elimination of the eave, • elimination of balconies, • changing column dimensions.	2; • changing place of balustrades on eaves, • the need for retrofitting.	• Contractor with owner, • Contractor with controller who is not the architect of the project.	He said that he could remember only the negative cases.
C4: Architect for 17 years, has his own office, the head of Chamber of Architects in North Cyprus.	7 changes in different cases.	1; • change in the width of staircase.	4; • suspended ceiling detail, -plinth detail, detail for division element between balconies, • change in the width of staircase.	3; • suspended ceiling detail, • detail of water channel in slope in wet spaces, • solution to the change in the width of staircase.	We mainly talked about the changes caused by master builders.	He said that the quality of application projects has been decreased in terms of details. He highlighted construction of green buildings with certificate.

project was childish." The architect managed to convince the client to proceed with the rework. Therefore, changes to the structural system must be overseen by civil engineers. However, while the components of structural, mechanical, and electrical systems are generally consistent and easy to identify, the factors that influence architecture can vary from building to building.

These four interviews revealed that numerous changes occurred in many of the cases they were involved in, though they did not examine these cases in detail. Table 8.6 presents the results of these interviews.

These four interviewees believe that decision making authority of contractors, builders, foremen, technicians, and carpenters about the tectonic process varies depending on the scale and complexity of the project. In larger and complex projects, specialized contractors who employ professionals in their teams can make significant contributions to these processes. However, in such cases, builders and foremen typically cannot make decisions. Their role is to alert professionals, who are responsible for making the final decisions. In contrast, for small and simple projects, the process is more flexible for these builders and foremen to make some decisions independently in order to keep the construction moving forward. Despite supporting the existing system, none of the interviewees endorsed the idea of builders and foremen making decisions about changes to the building. They were concerned that such changes could negatively impact the engineering systems or the architecture of the building.

Conclusion

Based on this chapter, which presents the overall results of this research, the main findings are as follows:

- As a case of abduction, the case of Monarga House demonstrates that *tectonics can be determined throughout the building process, not only during the architectural or application project.* The results of seven interviews support this approach to tectonics.
- Many actors can play significant roles in the process of tectonics. The results of the four interviews with experienced professionals active in construction sites contradict the RONAC's approach, as *all four interviewees advised against making changes without consulting professionals.*
- The construction process of the Monarga House (including the application project, tendering process, and Phases A, B, C, and D) can be related to many theories, including various philosophies (such as ethics, aesthetics, and politics), the theory of architecture, tectonics, and construction. It is also important to remember that many architectural theories, such as the Arts and

Crafts Movement and Bauhaus, emphasize the significant roles that various actors, such as craftsmen, can play in the tectonic process. The RONAC prefers to reconsider the relationship between professionals and builders, foremen, technicians, and carpenters in this context.

It is also possible to extend these findings and further contribute to tectonic theories by utilizing the rich categories of tectonic concepts presented in this book (such as types of errors and changes in hidden elements and components) to develop a theory of the "tectonics of details." This would acknowledge the contributions of builders, foremen, technicians, and carpenters to the tectonic process. Would it be possible to develop an argument on this issue?

9

Conclusion

Towards a Theory of Tectonics as a Process

The primary goal of this book is to investigate the influence of the building process on the tectonics of buildings and to establish a foundation for a theory of tectonics as a process. While one might assume that achieving this goal could be realized through questionnaires and detailed interviews with various individuals involved in building activities, the complexities of building processes present challenges. Building processes are complex, often involving numerous changes that may be revised multiple times. Many changes are interconnected, triggering cascading effects. For instance, consider the changes made in the heat insulation layer on reinforced concrete elements during the construction of the Monarga House. These changes are linked to the processes of constructing the reinforced concrete frame system, placing brick walls, applying heat insulation and adding plaster layers, including the leaf plaster. Consequently, various worker groups, including those involved in formwork, concrete pouring, brick laying, and plaster application, influence these processes.

Conducting interviews with foremen and workers might not help because they may not be aware of the relationships between the changes they implement. Controllers of buildings usually step in just before payments that occur at the end of each construction stage or at predetermined critical times. Despite knowing the expected outcomes, controllers may not be aware of changes that have occurred more than once. Additionally, recalling 175 changes, as in the case of the Monarga House process over a span of 2.5 years, can be challenging. Even if a series of interviews is conducted with controllers, questioning each stage separately, remembering each of changes and their relationships is difficult. If controllers were asked to record the changes, they would likely capture the significant ones rather than all of them. They might not be aware of the relationship of these changes to tectonic affects. To address this challenge, having one person involved in all roles – Researcher, Owner, Neighbor, Architect, Controller (the RONAC) – throughout the entire process of a building can be immensely

---

Tectonics as a Process in Architecture, First Edition. Yonca Hurol.
© 2025 John Wiley & Sons, Inc. Published 2025 by John Wiley & Sons, Inc.

280 | *9 Conclusion*

beneficial. This unique situation occurred during the building process of the Monarga House, allowing for a comprehensive collection of visual and written data on the maximum number of changes.

Objective Results of this Research

The hypothesis of this research is "tectonics is, or can be, a process." In Chapter 8, the culmination of this study shows that the building processes of the Monarga House and three other cases (OAC, Owner Architect, Controller; ONAC, Owner, Neighbor, Architect, Controller; and OCCC's buildings, Owner, Civil Engineer, Contractor, Controller), underwent numerous changes, with half of these changes significantly influencing the tectonics and even the architecture of the buildings. Additionally, the four professionals (one architect contractor, one architect controller, one civil engineer controller, and the head of the Chamber of Architects in North Cyprus) acknowledged that many changes are likely to occur during construction. Some of these changes, particularly in the Monarga House and Case 1 (because of the removal of the pitched roof), even resulted in revisions to the building's design decisions. Furthermore, the presence of change management literature as well as rework management literature indicate that changes are occurring in building processes worldwide. Meanwhile, caricatures depicting the negative effects of changes during construction demonstrate the substantial impact such changes can have on the tectonics and architecture of buildings. Consequently, *the results of this abductive case-study research and the seven interviews suggest a generalization that tectonics "is/can be" a process.* Just like the literature on change and rework management, all interviewees agreed that changes are not avoidable during building processes. However, changes may occur more frequently in North Cyprus due to its unique political and economic context.

Notably, various changes impacted the tectonics of the Monarga House both positively and negatively, showing that these modifications were not solely technical but also affected the meaning and aesthetics of the building. Engaging in theoretical reflections on philosophy, encompassing ethics, politics, aesthetics, and architectural theory, alongside technical theories such as construction and structural systems, throughout all chapters on the building process of the Monarga House, serves as additional evidence that tectonics is both an aesthetic and technical category simultaneously and it usually emerges as an outcome of a process. Additionally, some changes in the Monarga House resulted in innovative attitudes, suggesting that *treating tectonics as a process "can be" beneficial for enhancing tectonic qualities and fostering innovative attitudes*, which also affect the quality of the building. The results of the seven interviews also supported this finding. All interviewees agreed that making changes can be beneficial.

The OCCC from Case 2 stated, "Problems can be solved according to the available facilities on the spot." Similarly, the interviewed contractor mentioned "There are also changes we can make to align as closely as possible with the project."

Several changes in the Monarga House had no impact on tectonic affects, because these were hidden changes. However, some concealed changes, which do not adhere to standards or specifications, can influence the building's quality and potentially lead to negative tectonic affects due to the deterioration of its components. The two interviewees, OAC and OCCC, explained two interesting hidden details that arose from certain errors. Although these details are hidden, they cause tectonic affects and are also the result of innovative attitudes. One issue was related to a drainage problem affecting terrace ceramics, while the other involved a formwork error that was turned into an opportunity. Consequently, it is reasonable to generalize *that hidden aspects of buildings can also influence their tectonics. This consideration opens up the possibility of viewing building quality as a tectonic category.* Such a discussion might lead to a critique of tectonic theories that overlooked hidden details.

The application project phase of the Monarga House introduced ambiguities that potentially foster innovative attitudes. One of these concerned the application of heat insulation and leaf plaster together. Case 1 of OAC experienced a significant change during the application project. Although the roof design was a key element in design, the civil engineer suggested using a typical pitched roof for economic reasons, which the owners accepted. They later removed the pitched roof as it did not align with their original idea. However, this problem could have generated an innovative idea towards designing an inexpensive roof which is also aligned with the design idea. According to the existing literature,[1] it can be suggested that *the application project phase of buildings harbor considerable potential for changes and innovative attitudes during the building process.* This research revealed three differing perspectives about this issue.

The first perspective suggests that introducing some ambiguity in the application project and proposing alternative solutions can be beneficial. The map of innovative attitudes and the related table in Chapter 8 show that, many innovative attitudes occurred in parallel with tectonic affects, particularly with multiple changes driven by the application project, and tendering processes of the Monarga House. This suggests that these processes in building production can trigger not only multiple changes related to tectonic affects but also foster innovative attitudes. The second approach, initiated by the architect controller, suggests that application projects in North Cyprus have become less detailed and more ambiguous, which poses a problem due to the reduced presence

---

1 Mahairi McVicar, *Precision in Architecture – Certainty, Ambiguity and Deviation.* (Routledge, 2019).

of professionals on site. This interviewee sees the unplanned ambiguity in application projects as a problem. The third approach, put forth by both the contractor and the architect controller, highlights that large and/or special projects, such as those targeting sustainability certifications (e.g., LEED, Leadership in Energy and Environmental Design or BREEAM, Building Research Establishment Environmental Assessment Method), design-build projects, high-rise buildings, iconic projects, and mega urban developments, typically involve a contractor with an on-site office staffed by experienced professionals. In these technically complex projects, where collaboration is essential, many details are designed on site, and innovative changes are welcomed. This suggests that in this type of project the development of the application project runs parallel to the construction process.

Innovative attitudes during the tendering process of the Monarga House proved effective, influencing later changes and contributing to the tectonic affects of the building. This is one of the factors which enabled the building team to remain influential during the later stages of the construction process. The changes initiated by contractors, builders, and foremen had both positive and negative tectonic affects. However, if the four interviews with professionals are considered, *there are differing views on the generalization that contractors, sub-contractors, builders, and foremen "can" contribute to changes that result in positive tectonic affects and innovations.* One perspective suggests that recognizing tectonics as a process represents a more inclusive approach, emphasizing that buildings are not solely designed by professionals through architectural and application projects. The building process itself with all its actors can also influence, and even determine, the tectonics of buildings. However, according to the interviewed contractor and the two controllers, who were both experienced professionals, contractors, builders, and foremen typically make certain types of changes only during the construction of small buildings. These changes can be either beneficial or harmful, as they are often made with the goal of completing the work on time. However, these interviewees would prefer contractors, builders, and foremen to make decisions after consulting professionals. The head of Chamber of Architects in North Cyprus noted that, according to law, contractors are required to employ professionals, but this does not apply to builders. He emphasized that the presence of professionals on-site facilitates better communication with the controller. However, since contractors often take on more projects than they can effectively manage, their professionals may not be continuously on site. This factor puts pressure on builders and foremen on site to make certain decisions.

OCCC stated, "During the construction of large or sophisticated projects, such as retrofitting projects, sub-contractors/builders/foremen may raise issues, but the decision-makers should always be professionals." The interviewed contractor, who works on large projects with on-site offices and specialized agreements, added that foremen usually report problems to civil engineers and architects, who then address the issues. The professional team develops solutions and instructs the builders and foremen on how to proceed. OCCC gave an example of a change that could not be made by builders and foremen alone. This was a structural change involving the addition of a long-span slab (around 10–15 meters) into a gallery, highlighting the need for consulting civil engineers on such structural system changes.

The head of the Chamber of Architects in North Cyprus shared an example of a change that resulted in the destruction of the architectural character of a building. This occurred when a builder changed the position of the roof balustrades of an apartment without consulting the project architect. This event illustrates that decisions affecting a building's design are architectural matters and must be made by the architect. However, changes affecting architecture vary from building to building, whereas engineering systems (such as the structural system) are easier to identify.

On the other hand, the experience of the Monarga House demonstrates that there can be changes which do not impact the engineering systems, building code requirements, the architectural concept of the building, or the contractor in terms of cost and time. There were 52 changes within this category that could have been, or were, made by builders, foremen, technicians, and carpenters at the Monarga House. For example, issues such as the height of columns, and the positions of windows could have been identified by builders and foremen, if they had been better informed about the project. These types of changes, which are usually about mistakes and improvements, are marked with hashtags in Chapters 4–7. Therefore, collaborating with builders and foremen can still be beneficial especially when there is a lack of site supervision.

Potential Biases in This Research and How to Avoid Them

Before going further to develop a basis for a procedural tectonics theory, it is better to outline the ways the potential bias has been avoided within this research because the abductive case study methodology carries a heightened risk of bias due to the researcher's subjectivity.[2] To mitigate this, the first step is identifying potential sources of subjectivity. Understanding the researcher's role, in this case, the RONAC, and her relationship with the Monarga House is crucial. As stated in the

---

2 Rolf Johanson, "On Case Study Methodology." In *Methodologies in Housing Research*. (The Urban International Press, 2005): 30–39; Frank Conaty. "Abduction as a Methodological Approach to Case Study Research in Management Accounting – An Illustrative Case" *Accounting, Finance & Governance Review* 27, accessed November, 6, 2023. https://doi.org/10 .52399/001c.22171; Anna Dubois and Lars-Eric Gadde "Systematic Combining: An Abductive Approach to Case Research" *Journal of Business Research* 55, 2002: 553–560.

9 Conclusion

book's introduction, the RONAC is an academic architect specializing in tectonics, and the Monarga House is her own design and residence. Additionally, bias could stem from the RONAC's past issues with the initial contractor, influencing her subjective position towards changes before and after parting ways. In anticipation of these biases, the RONAC carefully reviewed and revised potentially problematic sections of the research results and theoretical contributions.

The second strategy to counter bias involves providing detailed contextual explanations. This serves to distinguish normal occurrences within the context from potentially exaggerated accounts, thereby minimizing sources of bias.

The third and primary strategy focuses on the inductive approach of the intermediate analytical chapters (3–7) in the book. These chapters follow an analytical approach, systematically presenting the changes made during each phase of the building process, their tectonic affects, and the associated innovative attitudes as factual. Additionally, the feelings of the RONAC are presented as psychological facts.[3] This qualitative but inductive method facilitates the objective validation of the hypothesis and the derivation of results through a structured and methodical process based on physical and psychological facts.

The inclusion of seven interviews also provides a fourth strategy to mitigate bias in this research. Some interview findings support the results of this study, while others offer different perspectives, enriching the overall outcomes.

Alternative Theory Fragments About Procedural Tectonics

Through the analysis of the theoretical reflections in Chapters 2–7, several alternative theory fragments have emerged, offering different pathways for parallel development alongside the proven hypothesis. These fragments encompass various critiques and concepts within tectonic theories:

- Critiques of existing tectonic theories:
  - The procedural character of tectonics (whether the changes are done by professionals or by contractors/builders/foremen or by professionals) is often ignored, treating buildings as complete frozen objects rather than processes. This also manifests in the insufficient emphasis on architectural and tectonic details within theories of architecture and tectonics.
  - The interaction between the concepts of architecture and tectonics has not been sufficiently addressed in theories of tectonics and architecture. The interviewees in this research believe that tectonics is more technical,

---

3 Edward Schippa and John P. Nordin, *Keeping Faith with Reason – A Theory of Practical Argumentation*. (Pearson, 2013).

while architecture is about functions and aesthetics. However, this distinction is not particularly helpful.

- Inapplicability of certain old concepts of tectonics, such as kernform, kunstform, and atectonic, to many contemporary buildings with modern systems, because these systems contain hidden layers.
- Not considering hidden parts of buildings in many tectonic theories causes problems.
- Challenging the judgmental nature of certain tectonic theories, particularly those addressing hidden building parts, by emphasizing the perspective of viewing the building as a process rather than just an outcome, is needed.
- Questioning the validity of tectonic judgments in the presence of "montage" and combining different approaches in architecture and tectonics, is necessary.

- Concepts for further development within tectonic theories: These are details, errors, and concealed/hidden building components/parts.
  - About details: System details, structure details, construction details, typical details, specially designed details, adapted details, functional and ornamental details, material and formal joints, being overdetailed, fertile details (leitmotif), and negative joints.
  - About errors: Concealed error, disturbing error, haptic error, benign error, errors to articulate, and fragile imperfection.
  - About concealing building components: Temporary or permanent camouflage, atectonic concealing, covering and painting, and making building parts visible.

- *Complementary concepts for a procedural tectonic theory: These are practicality, instrumentalization, and precision or imprecision in addition to innovation and innovative attitudes.*
  - About practicality: The typical, simple or complicated, practicality through technical improvement, practicality through innovation, and practical structure, form, building system, construction method.
  - About instrumentalization of building processes: Selecting a cheap but problematic site, ignoring architectural values except economy during design, not following the project and minimizing excavation or height/length of elements or components, the use of inappropriate materials and systems to reduce the cost and making the outcome look much better than it really is.
  - About innovation and innovative attitudes: Changing materials, systems, methods, and details, adapting unrelated objects to buildings (such as using a bird cage for the chimney cap to stop birds coming in), canceling something (such as canceling the stairway going to attic and replacing it with a simple trap door), changing for a better tectonics, identifying, and solving problems (such as eliminating dangers, e.g., drainage dangers), eliminating mistakes

or forgotten issues, being innovative and creative simultaneously and having and innovative attitude during preliminary design and application project and tendering process, and construction.

– About precision: Mathematical perfection, precision, placement of elements and components approximately, tolerance, order in nature, imprecision or irregularity, and errors. The concept of imprecision apparently relates to the various levels of the concept of tectonic error.

By considering the relationship between concepts like error, concealment, practicality and precision and imprecision, it is synthesized under the heading of detail. *Detail is the locus of tectonic processes.* A procedural tectonic theory, aligned with the hypothesis in this book that "tectonics is and *can be* a process," can be constructed by merging these alternative theory fragments. This inclusive approach allows for a holistic understanding of tectonics as a dynamic and evolving process in architecture.

On a Procedural Theory of Tectonics

A Critique of Some Tectonic Theories

While tectonics in architecture is an aesthetic category,[4] it is equally intertwined with building technology. Unlike aesthetic objects which are often treated as outcomes of artistic work, the consideration of building technologies (including materials, structural and construction systems, details, mechanical systems, electrical systems, etc.) demands an examination of processes. Overlooking the procedural nature of tectonics can negatively affect the essence of tectonic theories. This means that theories of architecture, tectonics and even construction are still influenced by the Ancient Greek concept of "aesthesis," which regarded objects as frozen in time and in their ideal forms. However, since the advent of Impressionism, modern art no longer views objects as static or in their best forms.

In addition, it can be said that the role of details in architectural and tectonic theories is insufficient. When considering the significance of details in the traditional or vernacular architecture, where they were viewed as the signature of the master-builder, this issue becomes even more obvious.

The interaction between the concepts of architecture and tectonics can be explored through the comments of the seven interviewees. ONACC stated "It is normal for changes during the construction of buildings to affect their tectonics,

---

4 Brian Kenneth Frampton, "Rappel a l'Ordre: The Case for the Tectonic." In *Labour, Work and Architecture*, Ed: K. Frampton. (Phaidon, 2002): 91–103; Brian Kenneth Frampton, *Studies in Tectonic Culture – The Poetics of Construction in Nineteenth and Twentieth Century Architecture.* (MIT Press, 1995).

On a Procedural Theory of Tectonics | **287**

but not their architecture." However, both the architecture and tectonics of his house were influenced by changes during the construction of the brutalist concrete walls in his project. Similarly, OAC noted that changes to the roof design of their house affected both its architecture and tectonics. These statements highlight the need for further discussion on distinguishing between architecture and tectonics.

The differentiation between the concepts of architecture and tectonics can be developed based on the types of architectural ideas and concepts. Some architectural ideas and concepts may be rooted in tectonic aspects, such as materials (such as Alvar Aalto's Villa Mairea), structural systems (such as Santiago Calatrava's Oriente Train Station), and details (such as Tadao Ando's Church of Light), while others may not. For example, the use of brutalist concrete walls was central to the architectural concept of ONACC's house. As a result, changes in the construction of the concrete walls affected both the architecture and tectonics of the building. Similarly, in OAC's initial house design, the repetitive triangles in the roof were a key architectural element. The radical simplification of the roof to decrease costs, and finally its removal, altered both the architecture and tectonics of the project.

This suggests that when an architectural idea or concept incorporates tectonic elements, changes in these tectonic aspects can also impact the overall architecture of the building. Furthermore, tectonics extends beyond details to encompass physical characteristics related to the use of light, wind, and topography, as well as cultural tectonic characteristics. This demonstrates the strong connection between architecture and tectonics, indicating that differentiating the two concepts is only feasible in specific cases and should not be generalized.

Many tectonic theories, akin to modernist architectural theories,[5] emphasize the *honest* (true) use of materials, structures, systems, and details through concepts, such as *kernform, atectonic,* and so on. However, certain concepts within these theories may not be applicable to contemporary buildings for two main reasons:

- Some of these concepts were developed concerning historical buildings with masonry structures in which structure usually forms the entire building. This makes them less relevant to contemporary buildings.
- Concepts for evaluating tectonic honesty may not be suitable for contemporary buildings with covered surfaces and hidden elements.

Concepts such as Karl Botticher's *kernform* and *kunstform,* Eduard Sekler's *atectonic,* and Kenneth Frampton's *tectonic form* face challenges in application to contemporary buildings with covered frames and hidden components. Kernform,

---

5 W. W. Braham and J. A. Hale (eds.), *Rethinking Architectural Technology – A Reader in Architectural Theory.* (Routledge, 2012[2007]).

9 Conclusion

for instance, is primarily relevant to structures with visible elements, as seen in stone Gothic cathedrals.[6] These concepts cannot even be applied to many Renaissance buildings, as they have *linaments* on their masonry walls. While linaments may look as if they are structural, they are not.[7]

Moreover, the value system of Eduard Sekler and Kenneth Frampton might not be applicable to many contemporary buildings, especially when tectonics is assessed only through completed buildings. Sekler's identification of hidden (illusory) parts as *atectonic*[8] and Frampton's call for making the construction process readable in buildings (through the concept of *tectonic form*)[9] are faint indicators of tectonics as a process, because they are read through the product rather than the building process.

Applying judgmental approaches to contemporary buildings can cast masterpieces, such as Le Corbusier's Ronchamp Chapel or Mies van der Rohe's Barcelona Pavilion, in a negative light due to difficulties in understanding hidden structural systems of the chapel, and chrome covered columns, and the presence of false walls in the pavilion, and so on.

Considering tectonic honesty is beneficial as it leads to a rationally equipped state of mind;[10] however, there are important architectural considerations in doing this:

- Approach tectonics as both a technical and aesthetic process to avoid overlooking potential problems and illusory situations. This avoids giving preference to either technology or aesthetics.
- Allow tolerance in evaluation, acknowledging that good architecture can encompass *errors* (such as sympathetic and benign errors, etc.),[11] *camouflage* (which is done to look better and develop better relations with other people),[12]

---

6 Karl Gottlieb Wilhelm Botticher, "The Principles of the Hellenic and Germanic Ways of Building with Regard to their Application to our Present Way of Building." In *What Style Should We Build?* Eds: J. Bloomfield, K. Forster and T. Reese. (The Getty Center Publication Programs, 1992[1828]): 147–168.

7 Gevork Hartoonian, *Ontology of Construction – On Nihilism of Technology and Theories of Modern Architecture.* (Cambridge University Press, 1994).

8 Eduard Sekler, "Structure, Construction, Tectonics." In *Structure in Art and Science*, Ed: G. Kepes. (1965): 89–95. accessed August 18, 2018. https://610f13.files.wordpress.com/2013/10/sekler_structure-construction-tectonics.pdf.

9 Frampton, "Rappel a l'Ordre: The Case for the Tectonic"; Frampton, *Studies in Tectonic Culture.*

10 Tim Ingold, *Making.* (Taylor and Francis-Routledge, 2013).

11 Alvar Aalto, "Inhimillinen Virhe (The Human Error)." In *Nain Puhui Alvar Aalto (Thus Spoke Alvar Aalto)*, Ed: G. Schildt. (Otava, 1997): 282; Alvar Aalto and Göran Schildt, "Speech at the Centennial Celebration of the Faculty of Architecture" on May 12, 1972 at Helsinki University of Technology. In *Alvar Aalto in his Own Words.* (Rizzoli, 1998).

12 Neil Leach, *Camouflage.* (MIT Press, 2006).

imperfections (such as in the fragile architecture),[13] and *montage* (which usually brings different tectonic value systems and styles side by side).[14]

- Evaluate each building example within its context, considering tectonics as a qualitative, rather than an absolute, judgment. Avoid developing overly judgmental theories about architecture and instead encourage critical questioning of ethical issues about specific buildings.
- Qualitatively assess each building critically, considering architects' perspectives[15] on tectonic, functional, environmental, and aesthetic values.

By adopting these considerations, discussions about tectonic honesty can be nuanced, contextual, and focused on the diversity inherent in contemporary architectural processes.

Actors of Procedural Tectonics: Recognizing the Role of All Actors in Building Processes

Many tectonic theories encounter challenges because they often consider buildings as completed objects, neglecting the procedural dimension which is vital for understanding the technical aspects of tectonics and the role of actors involved in these processes. Tectonic processes, as observed in the case of the Monarga House, led to changes. Some of these changes stemmed from innovative attitudes aimed at technical improvements and improved quality. Many of the changes result in pleasing or disturbing tectonic affects, as seen in the case of the Monarga House.

Although the majority of current building activities involve changes throughout the entire building process, this is not recognized in many tectonic/architectural

---

13 Juhani Pallasmaa "Hapticity and Time-Notes on Fragile Architecture" *Architectural Review* 207, 2000: 78–84. accessed September 3, 2020. https://pdfs.semanticscholar.org/e633/c06ae14c8fb9eeaadad27cde25432ac931ac.pdf. Gary Coates wrote that John Ruskin defended imperfection, irregularities and deficiencies in architecture. Gary Coates, *Erik Asmussen – Architect*. (Stockholm: Byggforlaget, 1997): 230; The concept of fragile architecture, characterized by its imperfection, serves as a reminder of the idea of "weak thought," Gevork Hartoonian, *Ontology of Construction*.
14 An example of merging different tectonic value systems is Adolf Loos' Looshaus in Vienna. This building features an explicit frame system on its lower floors, while the upper floors, though also framed, give the appearance of masonry. Thus, the frame system is used honestly on the lower floors but treated differently on the upper levels. This divergence is due to the required level of privacy in the hotel rooms. Gevork Hartoonian cites this building as an example of montage. Hartoonian, *Ontology of Construction*.
15 A critical approach requires evaluating others based on their own point of view. Theodor W Adorno, *Negative Dialectics*, Trans. E. B. Ashton. 2nd Edition. (Continuum, 1994[1966]); Hilde Heynen and Theodor W Adorno "Architecture between Modernity and Dwelling: Reflections on Adorno's Aesthetic Theory" *Assemblage – A Critical Journal of Architecture and Design Culture* 17, 1992: 84–91.

9 Conclusion

theories. Acknowledging tectonics as a process implies that design activity continues throughout the entire building process. According to the two main perspectives that emerged during this research, the procedural design activity either involves non-professional actors associated with the building (including contractors, builders, foremen, technicians, and carpenters) alongside the architects, engineers, and controllers, echoing principles of the Arts and Crafts Movement,[16] or it is a process that can only be determined by professionals. All professional interviewees in this study supported the second approach, providing clear examples of changes affecting the structural system and architecture negatively. Such changes could only be decided by professionals.

Can the oversight of contributions from contractors, sub-contractors, builders, foremen, technicians, and carpenters in design, innovation, and tectonic qualities be attributed to the *spontaneous ideology* prevalent within the professional realm, particularly among architects? Is design regarded as a sacred professional activity, with architects seen as the primary source of the aesthetic dimension?[17] Similarly, does the spontaneous ideology of engineering, focused on being scientific, sustainable, solving problems, and fostering innovation, also contribute to overlooking the input of contractors, builders, foremen, technicians, and carpenters? In this paradigm of spontaneous ideologies, are contractors and/or their team merely perceived as executors, tasked with producing physical buildings based on professional designs? At this point, it is important to differentiate between contractors, sub-contractors, builders, foremen, technicians and carpenters. Builders and foremen have less decision-making power over their work compared to the other groups.

While generalized answers to above questions are difficult to provide, it is clear that certain decisions, such as adding a slab to a gallery, are the domain of civil engineers, if that is accepted by the architect. The engineering design of that slab and its connection details should not be determined by anyone else. It can be generalized that changes in the structural system should always be discussed with a civil engineer, as civil engineering projects are based on specialized engineering knowledge. The dimensions of structural elements, the amount of reinforcement, or structural details should not be altered by anyone else. Changes to mechanical and electrical systems should also be discussed with the respective

---

16 Marco Frascari, "The Tell-the-tale Detail." In *The Building of Architecture*. (1984). accessed August 19, 2018. https://uwaterloo.ca/rome-program/sites/ca.rome-program/files/uploads/files/frascari-m-the-tell-the-tale-detail-3-a.pdf.

17 Louis Althusser, *Philosophy and the Spontaneous Philosophy of Scientists*. 2nd Edition. (Verso, 2011[1974]). Hasan Ünal Nalbantoğlu, "Yaratıcı Deha: Bir Modern Sanat Tabusunun Anatomisi (Creative Genius: The Anatomy of a Modern Art Taboo)." In *Çizgi Ötesinden: Üniversite: Sanat: Mimarlık*. (ODTÜ Mimarlık Fakültesi Yayınları, 2000): 85–104.

engineers, as these could impact both the comfort, safety, and value for money of the building. Similarly, the decision to change the position of the balustrade on a roof, as explained by the head of the Chamber of Architects, should be made by an architect, if it affects the conceptual unity of the building. On the other hand, decisions about other changes which may or may not impact the building's architecture vary depending on the specific building. Some changes, which are architecturally acceptable for the Monarga House, might not be acceptable for other buildings.

However, it is still possible to envision many changes initiated by experienced builders and foremen, provided there is a healthy collaboration between professionals, and builders and foremen are sufficiently informed about the specific building. The worldview on architecture, whether it is accepted as an art form created by the architect (as criticized through the concept of spontaneous ideology of architects) or as a functional and cultural utility that can be altered later by owners and users (as suggested by the Open Building theory), also influences the approach towards building production. This is also related to copyright of artistic works which suggests art works belong to their creators.

The head of the Chamber of Architects in North Cyprus stated that the relationship between the owner, contractor, and controller determines the quality of the outcome in North Cyprus. If the contractor makes decisions directly with the owner, the controller (who represents both the owner and the architect) is excluded from the process and this lowers the quality of the building.

As mentioned in Chapter 8, the changes that could directly be made by builders, foremen, technicians, and carpenters are those that:

- Do not require consultation with civil, mechanical, or electrical engineers
- Do not affect standards or building code requirements
- Do not conflict with the architectural concept, and
- Do not affect the contractor in terms of time or cost.

Of the 175 changes made during the construction of the Monarga House, 52 fell into this category. More than half of these had a positive or negative impact on the tectonic aspects of the project, and one-fifth showed potential for innovative approaches. Although less than half of these were successfully realized, this underscores that these individuals can also influence the tectonic process. It is also important to note that, historically, roles on construction sites were structured differently, with builders, foremen, and craftsmen playing a more prominent role in the past.

Architects collaborate with engineers to preserve the design intent of their buildings and prevent detrimental changes caused by engineering projects. Professionals also work closely with contractors, who are required to hire professionals in

9 Conclusion

North Cyprus, for the construction of large and special projects. Although all the professional interviewees in this research emphasized that no changes should be made by builders, foremen, technicians, and carpenters without consulting professionals, this study aims to demonstrate that consulting professionals is not always necessary for many changes. If experienced builders, foremen, technicians, and carpenters are informed about what they should generally avoid and are made aware of the specific architectural characteristics of the building, they can contribute more effectively to the construction process. They can prevent changes that negatively impact the architectural character of the building and implement supportive modifications that align with the architectural concept. The RONAC believes that these changes happen regardless of whether we acknowledge them or not. The architect controller also stated in his interview that foremen and workers sometimes make decisions independently to avoid losing time.

In North Cyprus, where on-site professional control is often limited, informing foremen about the architectural aspects of the project, enabling them to identify and implement certain changes, and ensuring they collaborate with controllers (who typically appear before payments are made) for other changes, can lead to positive tectonic developments and foster innovative attitudes. The holistic and inclusive approach not only enhances the tectonic qualities of buildings by encouraging contributions, such as innovation, from all actors and stakeholders, it also reduces the risk of misusing building processes. Many changes made by uninformed builders and foremen may seem like an instrumentalization of the building process. At the very least, such changes can be avoided.

Ethics and Building Production

The relationship of a project with its context[18] along with other values such as safety, functionality, aesthetics, and sustainability, forms the basis of architectural ethics. Professional rules in architecture combine deontological (duty) ethics,[19] virtue ethics,[20] and utilitarianism[21] to govern building processes[22] by defining the roles and responsibilities of architects within these processes. Many codes of conduct for architects emphasize legality, honesty, impartiality, equity, perfection, competence, integrity, and dignity. These codes outline the

---

18 Martin Heidegger, "Building Dwelling Thinking." In *Poetry, Language, Thought*. (Harper and Row, 2001): 147–148.

19 Immanuel Kant, *The Metaphysics of Morals*. (Cambridge University Press, 1996[1797]).

20 *Aristotle Nicomachean Ethics*. (University of Chicago Press, 2012[350 BCE]).

21 John S. Mill, "Utilitarianism." In *Collected Works of John Stuart Mill*, Ed: J. M. Robson. (Routledge and University of Toronto Press, 1991).

22 Thomas Fisher, *Architecture Design and Ethics: Tools for Survival*. (Routledge, 2018).

On a Procedural Theory of Tectonics | **293**

obligations of architects to the public, clients, the profession, colleagues, and professional organizations.[23] Architectural ethics also considers the importance of collaboration with other professionals (but not with all stakeholders such as builders and foremen) involved in building activities. Professional ethics can vary between different professions,[24] as the values specific to each profession may differ.

Legal building processes vary more significantly between countries compared to ethical principles. The phases of project preparation, permission for building, construction oversight, and final approval for use are all grounded in professional activities. The laws and rules which relate to contractors, sub-contractors, and builders usually concern the agreements and contracts they make with the owners of buildings, and the ethics that bind them is the *agreement ethics*.[25] These contracts between owners and contractors/builders define the legal roles and responsibilities of these parties. These agreements can take various forms, ranging from strictly defined written contracts to more flexible ones. Violations of contracts, which go against the ethics of agreements, can lead to legal disputes.

Issues such as contractors, sub-contractors, and builders, as well as professionals, taking on more projects than they can manage, resulting in the absence of regular professional control on construction sites (from either the contractors' side or as controllers), the long-term devaluation of professional knowledge about buildings (which usually becomes evident only after serious earthquakes), and the protection of the building sector through legislative, judicial (laws protect contractors in TRNC), and executive powers, should be understood as characteristics of the macro-political context in North Cyprus. In this context, where the construction sector has become the driving force of economic development, it has led to the absence of detailed application projects, a lack of control on site, frequent changes during the building process (often resulting in negative tectonic affects),

---

23 Union International des Architects, UIA Accord on Recommended International Standards of Professionalism in Architectural Practice, (2002–2017), accessed February 22, 2025. https://www.uia-architectes.org/wp-content/uploads/2022/03/uia_accord___updated_2017.pdf; American Institute of Architects (AIA), Code of Ethics and Professional Conduct, 2017, accessed February 22, 2025. https://cdn.ymaws.com/www.aiaqueensny.org/resource/resmgr/files/code_of_ethics__2017_.pdf; Architect Council of Europe-(ACE), European Deontological Code for Providers of Architectural Services, 2016, accessed February 22, 2025. https://www.ace-cae.eu/uploads/tx_jidocumentsview/Deont-code_2016.pdf; Union of Chambers of Turkish Civil Engineers and Architects (UCTEA), Chamber of Architects of Turkey (CAT), Codes in the sphere of Architecture, 2008, accessed February 22, 2025. https://www.tmmob.org.tr/etkinlik/muhendislik-mimarlik-kurultayi-2003/kurultay-kararlari-mesleki-davranis-ilkeleri; Seyit Ermiyagil, *Evaluation of Ethical Concerns at Architectural Drawings in North Cyprus*. (PhD Thesis, Eastern Mediterranean University, 2021).
24 Émile Durkheim, *Professional Ethics and Civic Morals*. (Routledge, 2018[1957]).
25 John Rawls, *A Theory of Justice*. 2nd Edition. (Harvard University Press, 1999[1971]).

a sharp rise in land and property prices, the instrumentalization of construction, and a preference for dominant technologies and typical details. This has contributed to affirmative architectural design rather than innovation, the adoption of a prescriptive building code instead of a performance based one, and the use of inexperienced builders, technicians, carpenters, and workers on construction sites.

This approach to legal issues and ethics in building processes often causes professionals to adopt a more defensive position, leaving little room for non-professional actors. The possibility of inclusive design, which involves the participation of contractors, builders, foremen, technicians, and carpenters, becomes even more limited in this framework. Unfortunately, this non-inclusive approach may reduce the enthusiasm of many non-professional actors in their roles.

Reasons for Change in Building Processes

There are approximately 50 factors that cause rework on construction sites,[26] and numerous factors led to changes during the building process of the Monarga House, including;

- Enhancing the commodity value of the building
- Resolving ambiguities in the application project
- Utilizing better materials, systems, and details
- Affordability considerations
- Addressing the availability problem of certain materials and systems
- Solving functional problems, such as optimizing window height for better ventilation in the Monarga House
- Avoiding low-quality construction
- Compensating for mistakes
- Changes triggering other changes in a cascading effect
- Owners' demands for changes and alterations
- Experienced contractors, well-versed in the local building market, may propose highly innovative changes during the tendering process to impress owners and secure the project.

Examples of these reasons for change in building process of the Monarga House are detailed in Chapters 3–7 of this book, demonstrating the multifaceted nature of changes in architecture.

---

26 Ramin Asadi et al., "Analyzing Underlying Factors of Rework in Generating Conractual Claims in Construction Projects" *Journal of Construction and Engineering Management* 149, no. 6, 2023: 04023036.

Beneficial and Harmful Tectonic Processes: A Materialistic Approach to Architecture

Recognizing tectonics as a process within tectonics theories brings forth the capacity to evaluate the positive and negative dimensions resulting from the procedural nature of tectonics. Tectonic processes can be beneficial or instrumentalized. This acknowledgment facilitates understanding of the positive or negative impacts of tectonic processes on various tectonic issues, such as errors and details.

Therefore, a procedural tectonics theory should encompass the following key issues:

- Beneficial tectonic processes
- Instrumentalized tectonic processes
- Characteristic issues of tectonic processes, such as errors, details, camouflage, montage, practicality, precision and imprecision.

Making Tectonic Processes Beneficial

The building of the Monarga House serves as a case study demonstrating that procedural changes can lead to positive or negative tectonic affects and innovative ideas. Although the RONAC was initially resistant to changes, she witnessed various changes contributing to tectonic qualities of the building and some changes were unavoidable. Examples of positive changes include the extensive use of timber, the incorporation of stone walls in the garden, the addition of rain chains, and the optimal positioning of the fireplace. This abductive case study and the seven interviews underscore the potential for improving building tectonics through changes made during the building process.

Once it becomes possible to recognize that tectonics can be (or even is) a process, it is worthwhile to explore how we can improve its benefits to achieve better quality tectonics. This is not merely a matter of opening buildings to change throughout the entire lifespan, as seen in the case of Open Building theories.[27] These theories encompass the entire design, production, and utilization processes of buildings with the goal of achieving adaptability and flexibility. It is also not about pursuing a dramatic innovative process, as exemplified by the Beijing Water Cube.[28] The Beijing Water Cube was conceived with an unknown structure and façade materials in the architectural competition project. Its feasibility was realized by incorporating specific innovative firms into the design process. This book explores changes and innovative attitudes aimed at improving the tectonic quality of simple buildings. The objective is not *change for the sake of change*, nor is it *innovation for the sake of innovation*. However, this approach is not modest, as a pursuit of

---

27 John Habraken, *The Structure of the Ordinary*. (MIT Press, 2000).
28 Anon, Water Cube – Beijing, accessed September 24, 2023. https://www.youtube.com/watch?v=CTUZe57ONvk.

9 Conclusion

better tectonics through change, innovative attitudes and more inclusive building processes, because it encompasses the majority of buildings and their building processes.

A professional attitude predominates legal building production in many countries, but the laws governing business agreements, professional rules, regulations, and codes of conduct vary across nations. However, in many countries, such as North Cyprus, professionals are unable to adequately control building processes. This section of the book seeks to scrutinize the entire legal building processes and rules, with a particular focus on their inclusivity to provide a solution for this problem. Exploring ways to improve inclusivity in building processes could prove beneficial from the standpoint of tectonics too. There is also a technological foundation for questioning the inclusivity of building processes that facilitate more change and innovation to achieve improved tectonics. Advances in building information management (BIM), its emerging technologies, and artificial intelligence (AI) have the potential to support and enable changes and innovations during the building process, even for small buildings, by facilitating a parallel and inclusive design approach.

Interestingly, the eminent pioneering civil engineer Thomas Telford designed and constructed his bridge projects by modifying his designs while building them. This was a practice he referred as "one-to-one modelling."[29] Amos Rapoport also advocated for continuous change in all processes and the lifespan of buildings, considering the changes within their contexts.[30] Additionally, the current contracting approaches, that involve having an on-site office, further demonstrate the benefits of being open to changes during the construction process.

Although this subject requires further research, the types of changes made during the building process of the Monarga House can provide insight into how building processes can be more inclusive. These include:

- The application project of the Monarga House contained some ambiguities, which led to a series of changes during the later building process. Ambiguities such as the relationship between the heat insulation layer and leaf plaster could have led to innovative attitudes and contributions to tectonics later, but they have not.
- The uncommon materials and systems that might lead to low-quality construction were changed during the application project phase. The cancelation of accordion windows and sliding shutters is an example of this. These also could have led to innovative attitudes and contributions to tectonics later, but they have not.

---

29 David P Billington, *The Tower and Bridge*. (Basic Books, 1983).
30 Amos Rapoport, *Culture, Architecture and Design*. (Locke Science Publishing Company Ltd., 2005).

On a Procedural Theory of Tectonics | 297

- The tendering process was beneficial because some of the contractors had some innovative attitudes to persuade the owner to offer them the job. They identified problems in the application project and suggested solutions for them. They also understood the tectonics of the project and suggested better materials, details, and so on. An example of this is the use of good-quality timber surfaces under the eaves of the roof. Such changes also initiated a series of changes and innovative attitudes in the later phases and improved the tectonic qualities of the building.
- Several reasonable changes were made or suggested by foremen, technicians, and carpenters, including adjusting the height of the house windows to improve ventilation, reducing the extension of pergolas, adding small transom windows above interior door frames, and resolving the angle issue of the garden's stone wall.
- Some changes were made to compensate for some mistakes that happened during construction. For example, when the application of the heat insulation layer on the reinforced concrete elements became impossible, the main controller and the RONAC negotiated with the contractor and asked for timber interior door frames. This was accepted.
- Unavailability or impracticality of some materials and systems also provide an opportunity to make contributive changes. Changing the garden walls from brick to stone was in this category. The contractor did not want to have brick walls with three layers of plaster on them. It was more practical to have stone walls without plaster. This was accepted by the RONAC.

In this book, the actors of change were categorized into five main groups: the architect of the application project, contractors, sub-contractors, builders, foremen, technicians, carpenters, and workers (the contractor's side), experienced actors such as builders, foremen, technicians, and carpenters (within contractor's side), the experienced ones who have less decision-making power such as builders and foremen (also within contractor's side) and the main controller and the RONAC (the owner's side). The distinction between contractor, sub-contractor, builder, foremen, technician, carpenter, and worker posed some difficulties, primarily due to the shifting nature of their roles. Workers and foremen eventually became builders. Also, these actors were on the same side; there was the contractor's side and the owner's side during the construction process. Sub-contractors usually produce certain components and systems such as interior doors, wardrobes and cupboards and aluminum windows and doors, and implement them, while contractors organize the construction of the entire building. Many technicians work for sub-contractors, and they can be specialist carpenters, mechanics, or electricians. Builders usually construct certain parts of buildings, such as the reinforced concrete frame, earthwork and timber roofs. Foremen and workers work for contractors or builders.

However, it can be expressed here that the maximum number of contributions came from the contractor's side during the process of creating the Monarga House. The contractor made a significant contribution, followed by considerable input from the experienced builders, foremen, technicians, and carpenters. A foreman stated, "We rarely realize that there is a mistake in the project. An experienced foremen consults with professionals on such matters. For example, there can be issues with angles or plastering problems due to the quality of the formwork, or if the formwork is difficult to prepare." This expression shows that although foremen made many decisions about details, they do not accept or declare these as changes. Workers contributed less because most of them were young, foreign migrants, and preferred to avoid communication with the RONAC and the main controller. Change management techniques, such as change orders, were not used during the construction of the Monarga House.

There can be systematic approaches to involve contractors, sub-contractors, builders, foremen, technicians, and carpenters, more effectively, in the building process. As seen in some factories and production plants,[31] workers and professionals could be organized into a network (within a rhizomatic structure) rather than a traditional hierarchical system, encouraging contributions from all.[32] Collaboration with contractors, sub-contractors, technicians, carpenters, and experienced and capable builders, foremen, workers, could be designed to provide them with detailed architectural information for specific projects, allowing them to oversee changes on site. As a result of this approach, the application project phase, the tendering phase, and the construction phase might undergo considerable changes.

- The application project may contain systematic ambiguities with some alternative solutions: These might be clarified during the tendering process(es) through innovative attitudes that may produce different solutions. The changes made during the tendering process can be about the characteristics of the preliminary design project that have not been fully reached in the application project. As the architect controller interviewee complained, application projects have become less detailed in North Cyprus, and this is contrary to having planned ambiguities in application projects.

---

31 Krishnamurty Subramanian "Employee Contribution, Empowerment and Organizational Results" *International Research Journal of Advanced Engineering and Science* 2, no. 3, 2017: 16–20. accessed September 27, 2023. https://www.researchgate.net/publication/318055319_Employee_Contribution_Empowerment_and_Organizational_Results#fullTextFileContent; Diana Leon "From the Sustainable Organization to Sustainable Knowledge-based Organization" *Economic Insights – Trends and Challenges* II, no. 2, 2013: 63–73.
32 Gilles Deleuze and Felix Guattari, *A Thousand Plateaus, Capitalism and Schizophrenia*, Trans. Brian Massumi. 5th Edition. (Continuum, 2004[1980]); Gilles Deleuze and Felix Guattari, *Anti-Oedipus – Introduction to Schizoanalysis*, Trans: Eugene Holland. (Routledge, 2002[1972]).

On a Procedural Theory of Tectonics | 299

- *Any possibility of low-quality construction* should be considered carefully because the application of such building elements, materials, and systems should encourage innovative attitudes especially during the application project phase and the tendering process.
- The building process can be divided into pieces, and separate tendering processes can be realized for each: It is known that this is beneficial because contractors, sub-contractors, and builders use their innovative attitudes to influence the owners. They identify problems in the project and suggest solutions. If they are experienced professionals, they also understand the tectonics of the project and suggest better materials, details, and so on. If the number of such tendering processes increases, this can be beneficial for the tectonics of the building. This might require a different type of agreement between the contractor, sub-contractor, builder, and the owner.
- Collaboration with experienced foremen/technicians/carpenters: This issue has remained unresolved since the Arts and Crafts Movement. Collaboration with builders, foremen, technicians, and carpenters would allow for continuous on-site control, complementing the sequential control activities led by controllers. The four interviews with professionals suggest that foremen on large construction sites in North Cyprus are typically subjected to rigid hierarchical structure, limiting their capacity to make decisions. In contrast, they tend to have more influence and responsibility on smaller sites. This also depends on whether a contractor (who employs professionals) or various builders (without professionals) are active in producing that building. The RONAC believes that the hierarchical structure affecting especially builders and foremen can vary from country to country. In some contexts, builders and foremen may be granted more decision-making power, particularly in resolving day-to-day issues, such as material shortages, time management issues, and ambiguities about the details. This highlights that the potential benefit of redefining the foremen's roles, particularly in contexts where there is a high rate of change, labor trafficking, or insufficient control by professionals. A more flexible and case-specific approach could indeed enhance efficiency on both small and large construction sites by empowering experienced foremen to address issues that arise more directly.

Defining tectonics as a process and including contractors, sub-contractors, and builders as well as foremen, technicians, and carpenters to tectonic design is a *materialistic* approach. When compared with the current abstract, artistic (considering design principles and the importance of form in architecture), mathematical (considering engineering processes), and precise design of contemporary buildings, tectonics as a process is more materialistic. The outcomes of processes like this can be more haptic, poetic, and can also produce friendly and artistic errors, unique details, and so on. All types of details, including system details, structural details, construction details, and so on, gain more importance. The influence of hidden details might also change because the building starts having the character of a body with its visible and concealed parts. Including contractors, builders, foremen, technicians, and carpenters in tectonic design can open many new pathways for architecture. Some materialistic approaches to architecture, such as the minor architecture and non-patriarchal approaches to architecture, can be harmonious with this approach too. Although the basic architectural intention *to clear out* cannot be entirely avoided, and the changes depend on many factors, the relationship with nature can still improve.

However, without proper oversight, and if the contractors' side has corrupt intentions, building processes can easily be manipulated, in other terms instrumentalized. Unfortunately, this happens and causes many problems, especially in high-risk earthquake regions such as North Cyprus and Turkey.

Instrumentalized Tectonic Processes
Instrumentalization of architecture and construction is not ethical or legal. There are legal procedures, professional rules (such as building codes, regulations, and standards) and professional values to prevent instrumentalization, which leads to low-quality buildings and architecture. However, the devaluation of professional knowledge and values in the building sector could lead the risk of succumbing to instrumentalization. Such instrumentalized changes may be influenced by the macro-political context.

According to Chapter 2, the context of the Monarga House presents risks of ethical problems and corruption due to the broader political and economic issues at the macro level. This political-economic context might have contributed to the high number of changes in the processes related to the Monarga House. Many Cypriots believe that political economy has a significant impact on building activities in North Cyprus. Political economy in North Cyprus directs the majority of building activities towards an unpredictable situation, especially because of the changes in currency. Estimating the cost of buildings is not possible due to high inflation and rapid increase in the economic value of all types of property. This has been seriously affected by the pandemic that occurred after the construction of the Monarga House. Also, the future of properties in North Cyprus, a contested zone, is generally ambiguous. The economic value of houses is extremely high, regardless of the way people achieve them: by buying a ready-made house, by having it built by a contractor or by builders, by using any building technology (such as prefabricated, timber, steel, or reinforced concrete systems), it does not matter. Despite a contract designed to ensure a swift construction process, the building of the Monarga House progressed extremely slowly.

If professional values and business ethics are disregarded by any actor(s) within the building process merely for financial gain, this can be termed an *instrumental approach*[33] to architecture and/or building activity. All building processes, including architectural design process, can be instrumentalized. Instrumentalization can also be partial in any building process. Architectural and tectonic signs of an instrumental approach in buildings are known for being monotonous and having both technical and artistic low quality.[34] Instrumentalization can also lead to significant changes in buildings concerning architectural projects, and the building might even become unrecognizable.

Instrumentalization of building processes can occur through various stages, such as selecting a cheap and problematic site, disregarding architectural values in favor of economic considerations during design (e.g., placing houses too close together with no green spaces), providing insufficient detail in application projects, relying on dominant building technologies in that context, choosing familiar and cheaper materials, systems, and details, deviating from the original project and minimizing excavation or reducing the height of columns and walls, or other components, using inappropriate materials and systems, and making the final outcome appear better than it actually is.

Although the Monarga House was not fully designed and produced with instrumental intentions, its building process provided an opportunity to contemplate the processes of instrumentalization in building production. Excavating less than specified in the structural engineering project, making the columns and roof shorter than specified in the application project, using cheaper materials and systems, eliminating heat insulation, poorly producing hidden details in contrast to making any visible parts look good, ignoring future deterioration due to poor detailing in the roof, applying bad camouflage (attempting to use gypsum to cover gaps between walls and door frames, wardrobe frames, etc.) are typically poor practices observed during the processes of the Monarga House.

There can even be a *typical image of instrumentalized buildings* in the minds of some contractors, sub-contractors, and builders, and this might lead to significant

---

33 Martin Heidegger's concepts of enframing and standing reserve can be used to explain the concept of instrumental rationality. Enframing refers to the way beings are defined and understood in a particular manner, shaping how everyone perceives them. Standing reserve, on the other hand, describes how the essence of beings is reduced, making them exist solely for their future economic potential. Heidegger Martin, "The Question Concerning Technology." In *The Question Concerning Technology and Other Essays*, Trans. William Lovitt. (Garland Publishing, 1977[1954]a): 3–35; Martin Heidegger, "The Age of World Picture." In *The Question Concerning Technology and Other Essays*, Trans. William Lovitt. (Garland Publishing, 1977[1954]b): 115–154; Martin Heidegger, "Science and Reflection." In *The Question Concerning Technology and Other Essays*, Trans. William Lovitt. (Garland Publishing, 1977[1954]c): 155–182.
34 Habraken, *The Structure of the Ordinary*.

changes in the appearance of many buildings. The overall image of any building can be severely affected by certain changes, such as making the columns and roof shorter than suggested in the architectural project, and so on. Since the steps leading to instrumentalization are often similar within the same city, country, or context, it results in monotonous environments due to the repetition of the same problematic characteristics. The relationship between context and instrumentalization requires further research.

Hidden changes towards instrumentalization might also occur. There could be poor but concealed details that do not align with the application project. An example from the Monarga House is the elimination of one roof layer, which allows water entering between the roof layers to escape. This could have led to long-term deterioration of the roof, making it look bad and consequently affecting the tectonics of the building. This change was somehow avoided in the Monarga House, and the detail was applied similar to the one in the application project.

There might also be *bad camouflage* that can cause deterioration in the long term. Not all camouflage proves effective in practice. An example of this within the process of the Monarga House is the suggestion of using gypsum for any gap between the elements of construction by the sub-contractor firm. This was not accepted, and the problem had to be solved by changing the dimensions of some elements, or by using some additional elements.[35] In fact, the use of gypsum for such purposes does not align with the definition of camouflage. Neil Leach defines camouflage as a strategy for connecting people by making things appear different, with an emphasis on aesthetically pleasing applications. The applied solution (such as using wider cornices) in the Monarga House to avoid the use of gypsum can be categorized as camouflage because they were done with the intention of achieving a healthy and aesthetic solution that will produce a good image in the long term.

Giving more importance to the tectonics of visible parts and ignorance of the concealed and covered parts are not within the content of many tectonic theories because they do not consider the building process as a whole. However, these problems are common in the building practice. The water outlets at the edges of the roof head-beam of the Monarga House were built wrongly first, and the problem was half solved later. Also, there were problems in the application project with the water and heat insulation layers on the reinforced concrete slab in the cool roof. Solutions were suggested during the tendering process, but the problems about these concealed details were also half solved. Acknowledgment of the contributions of contractors, sub-contractors, and foremen might encourage these people to build the hidden parts of buildings better than specified in the application project. This acknowledgment might also motivate them to rectify

---

35 Neil Leach, *Camouflage*. (MIT Press, 2006).

On a Procedural Theory of Tectonics | **303**

errors in the application project. The failure to assign tectonic value to hidden parts of buildings should be highlighted as an ethical issue in tectonics. Giving importance only to the tectonics of visible elements results in a *seductive surface architecture*, which differs from the *surface architecture* defined by Leatherbarrow and Mostafavi.[36] Leatherbarrow and Mostafavi employed the concept of surface architecture to elucidate the aesthetic transformations occurring during modern times, and the concept did not carry such heavy negative connotations.

Prioritizing visible parts while ignoring hidden parts of buildings bears some resemblance to Adorno's definition of *fetishist detailing*.[37] Adorno was opposed to having either purely functional or purely aesthetic details. He advocated for details with both functional and aesthetic value to avoid fetishism. While there were visible details in the Monarga House that were both functional and aesthetic (such as the rain chains), the presence of nonfunctional and poorly designed concealed details (such as the poorly executed roof headboard details), renders these aspects fetishist. It might even be more apt to label this *post-fetishism*, as it pertains to giving importance to visible parts and disregarding the hidden parts of buildings.

There might be serious ethical problems arising from changes made within the building process by contractors, sub-contractors, builders, foremen, and workers leading to the instrumentalization of the process and a violation of the agreement between the owner and the contractor, sub-contractor, or builder. These actions contradict both *agreement ethics* and *virtue ethics*.[38] Such practices can easily result in unsafe buildings that may collapse during earthquakes, floods, tornados, and so on. The RONAC believes that the 2023 Turkish earthquake, which resulted in the deaths of 50,000 people,[39] was largely due to extensive instrumentalization in land selection, design projects, and building control processes which are defined by legal procedures.

The type of innovation expected from contractors, sub-contractors, builders, and foremen, within such corrupt processes can only lean towards instrumentalization. This may be termed *instrumentalization of innovative attitude*. The design of the water outlet details beside the fascia of the timber roof of the Monarga House can also be seen as an example of instrumentalized innovation. The roof was completed with this problem. As numerous alterations were required in the roof construction to address this issue, incurring substantial costs, the main controller and the RONAC opted for a slightly improved (though still suboptimal)

---

36 David Leatherbarrow and Mohsen Mostafavi, *Surface Architecture*. (MIT Press, 2005).
37 Theodor W Adorno "Functionalism Today" *Oppositions* 17, 1979: 31–41.
38 John Rawls, *A Theory of Justice; Aristotle Nicomachean Ethics*. (University of Chicago Press, 2012[350 bce]).
39 UNDP Türkiye, 100 Days After Earthquakes, May 17, 2023, accessed February 22, 2025. https://stories.undp.org/100-days-later#!.

9 Conclusion

solution. They bored small holes in timber surfaces to allow water drainage. It is noteworthy that this instrumentalized, innovative approach was initiated by the RONAC and the main controller.

Details as Loci of Tectonic Processes

Tectonic processes can influence the structure of the building, but it is unusual for them to result in changes in the form of the building. The process can cause changes in the materials used; however, it is usually not desired to have a considerable change in the general image of the building. On the other hand, tectonic processes cause many changes in details of buildings. They can cause changes in:

- System details
- Structure details
- Construction details.

The above three categories of details indicate the different drawing scales that can appear in the application projects. System details are drawn in a 1/20 scale, while structure and construction details, which usually express joints, are drawn in a 1/5 or 1/1 scale.

Details can be investigated through some other technical concepts as well. These are:

- Typical details, which are commonly used.
- Specially designed details, which are unique and specifically designed for that particular building. This can be an innovative process.
- Adapted details, which are taken from another source and changed according to climate, function, and so on. The adaptation of technologies requires adaptation of details. This is an innovative process too.

The above two categorizations of details demonstrate a technical approach. However, the tectonic character of details requires the use of other concepts. Louis Kahn praised tectonic details twice in his poems as "Ornament is adoration of the joint" and "Detail is the adoration of nature."[40] The concepts which represent tectonic character of details are:

- Functional/ornamental detail: Theodor Adorno suggested architects have both functional and aesthetic (ornamental) details. According to him, a detail that is only ornamental is fetishist,[41] and a detail that is only functional is not architectural.

---

40 Ian Volner, Sunsets: Scarpa's Tomba Brion, Kahn's Salk Institute, and Geomancy, accessed September 14, 2024. https://archive.pinupmagazine.org/articles/article-tomba-brion-salk-institute-sun-architecture-geomancy#48.
41 Adorno, "Functionalism Today."

On a Procedural Theory of Tectonics | 305

- Material and formal detail (joint): Marco Frascari wrote that formal details (joints) separate different functions in a building, such as separation of indoor and outdoor environments, but material details (joints) do not separate different functions.[42]
- Being specified as overdetailed is a critique about exaggeration regarding the number of details because some details are about signification, and they are expected to divert attention towards something meaningful. However, if there are too many significant details as such, the diversion of attention to a meaning might not be realized properly.[43]
- Fertile detail (like "*leitmotifs*" in classical music, such as in operas by Richard Wagner)[44]: Fertile details are repeated in one building, in many buildings by a particular architect, or they are repeated generally by other architects. They may even become characteristic of an environment. The ziggurat motif in Carlo Scarpa's architecture is an example of fertile details because he repeated this motif in his many buildings. On the other hand, Andrea Palladio's Serliana Window became the Palladian Window in time and is commonly used.[45]
- Negative detail (joint): If a detail is designed to give the impression that one part of the building does not exist, it is called a negative detail.[46] The corner windows in Carlo Scarpa's Extension to the Canoviano Museum in Possagno are examples of negative details because they give the impression that those inward windows do not exist, and their places are empty.[47]

Some details contain innovative attitudes of various actors in the building processes. These innovative attitudes can be about:

- Changing materials, systems, methods, and details, such as changing the material of garden walls from bricks to stone to avoid plastering processes in the Monarga House
- Adapting unrelated things to buildings, such as using a bird cage on the chimney cap to stop birds coming into the chimney,
- Canceling something, such as canceling the stairs to the attic and replacing it with a practical timber trap door
- Changing for better tectonics, such as using better quality timber under the eaves of the roof

---

42 Frascari, "The Tell-the-tale Detail."
43 Frascari, "The Tell-the-tale Detail."
44 Matthew Bribitzer-Stull, *Understanding the Leitmotif: From Wagner to Hollywood Film Music.* (Cambridge University Press, 2015).
45 Frascari, "The Tell-the-tale Detail."
46 Frascari, "The Tell-the-tale Detail."
47 Stephan Buzas and Judith Arthur, *Carlo Scarpa: Gipsoteca Canoviana, Possagno.* (National Book Network, Opus 22 Series, 2002).

9 Conclusion

- Identifying and solving problems, such as eliminating the danger of flooding by providing special drainage details
- Eliminating mistakes, such as solving the angle problem between the perpendicular patios at the entrance of the house and avoid having an unattractive arrangement of ceramic tiles in them
- Installing forgotten items, such as adding a septic pumper truck entrance to the garden.

These innovative attitudes might occur during the entire building process, including the application project, tendering process, and the construction phase. Some of them can also be creative by considering the aesthetic dimension simultaneously.

Different tectonic concepts and characteristics can be related to details. These are:

- *Concealing* building parts, whether they are bad or not, form details.
- *Errors* can be seen as details, and they can also trigger different details.
- Details can be *practical*, and they can make the whole process practical or impractical.
- Details can be *precise* (as in the case of the technical perfectionism of Mies van der Rohe) or imprecise (as in the case of fragile architecture) as well.

It is useful to collect all these concepts and characteristics under the umbrella of details, which has not been studied sufficiently before in tectonic theories.

*Camouflage*s are formed through construction details. They have small or large surfaces that cover and connect to other surfaces or elements. Building parts can be left visible, or they can be covered with painting, or with camouflage. Is painting a layer of camouflage? It is not because it is a protective layer, and it can also be an aesthetic layer. Camouflages can be temporary or permanent aesthetic covers. The following categories regarding camouflage are different from each other:

- Having no camouflage: having visible building parts
- Having plaster and painting as protective and aesthetic layers
- There can be permanent or temporary aesthetic covers
- Some camouflage attempts might cause technical problems in long term.

There can be different value systems to evaluate these categories. The first two types are very common applications. Neil Leach described the third one as an aesthetic action.[48] However, the fourth type is not acceptable.

Some of the changes that caused tectonic affects in the Monarga House can be useful in thinking about these categories. The detail about the reinforced concrete frame and eco-brick walls covered with leaf plaster can be useful in understanding

---

48 Leach, *Camouflage.*

On a Procedural Theory of Tectonics | **307**

the second category. Eco-brick walls are not complete if they are not plastered. Plaster on these walls helps keep bricks together, protects the wall, and it is also an aesthetic layer, while the leaf plaster is a special aesthetic type of plaster. The typical plaster layers inside the Monarga House cover and protect the cables and pipes on the walls and ceilings. It is not pleasant to see all those cables and pipes, unless that is the target of the project. These examples make us think about the second category.

The discussion about adding gypsum layers to close the gap between interior door frames and the cornices of those doors of the Monarga House can be useful for evaluating the last two types of camouflage. The actual function of cornices is to cover the gap between door frames and walls. Still, the builder said that gypsum can be used as a permanent aesthetic cover for the gap between the cornice and the wall. The main controller and the RONAC thought that use of gypsum there would cause cracks in the long term (the fourth type) and refused to use it. Finally, larger cornices were used for these doors, and the problem was eliminated. Larger gaps occurred between the wardrobes and walls as well, and the builder suggested using gypsum again. The main controller and the RONAC refused this too because it was again the fourth type of camouflage. Finally, useless wardrobe doors were added in front of those gaps as permanent camouflage for aesthetic purposes only (the third type). The judgments used in this last situation were about refusing short-term camouflage when a long-term camouflage was needed. The fourth type of camouflage can be called *atectonic concealing* because it is a short-term solution, and it will cause deterioration soon. The use of gypsum in these cases is hiding one's own error in a bad way. After the gypsum causes cracks, it is too late. These *badly concealed errors* become *disturbing errors* after they are cracked.

If cornices were sufficient to cover the gap between the door frame and the wall, these details could be called *typical construction details*. However, using a larger cornice to cover the gap turns it into a *specially designed construction detail*. These are *functional and material details*. The way the gap between the wall and the wardrobe closed is different because it *adapts another typical detail* to solve this problem by using a wardrobe door to cover the gap. This is a *material and ornamental detail* because that wardrobe door is too high, and it cannot be used easily. None of these details can be categorized as functional and ornamental simultaneously.[49] However, the use of leaf plaster as a protective and aesthetic layer can be categorized as a *functional and ornamental detail*.

The ONACC of Case 3, one of the interviewees in this research, mentioned an interesting case of camouflage. His architectural project featured brutalist concrete walls, which were conceptually significant. However, despite the contractor's assurance that he could construct high-quality brutalist concrete walls, he ended

---

49 Adorno, "Functionalism Today."

308 | *9 Conclusion*

up casting them piece by piece and using substandard formwork materials. As a result, the ONACC had to apply a special plaster to make these walls resemble the intended brutalist concrete. This plaster layer created an interesting camouflage, as it was employed to imitate the architectural characteristics originally specified in the design. The ONACC stated that, "Ordinary people would not realize that these walls are not brutalist concrete, but I know what happened, and I feel sad about it. I did my best with them."[50]

Some details can be erratic, and tectonic errors can be categorized as follows:

- Disturbing error, such as the cracked gypsum infill
- Haptic error
- Benign (or sympathetic) error
- Errors to articulate
- Fragile imperfection.

The RONAC can remember only one detail in the Monarga House that can be called a *haptic error*. It is such a small change that it has not been mentioned in this book before. Since the foundations of the house entrance patio and the walkway towards the house were built separately, there occurred height difference between them of 1 cm. This is like a door sill; however there is no door there, and people do not expect this little change in height, so they trip there. This error can work like a haptic error, which can cause memory flashbacks over time and remind people of their past memories if they visit the place in future and trip again at the same place.[51] This detail can also be categorized as a *material construction detail*, which is neither functional nor ornamental.

There are no benign errors[52] in the Monarga House. A *benign error* is a way of being critical about the demand for architectural perfection. Alvar Aalto's Villa Mairea contains such benign errors because Aalto intentionally created some irregularities in this building. Two examples of these are the different ways of wrapping the timber columns and the irregular spacing of the timber vertical elements around the staircase. Villa Mairea details are material details, and they are both functional and ornamental. There can also be unintentional benign errors, such as slightly curved marble panels of the Finlandia Hall of again, Alvar Aalto. Marbles became curved because of high temperature differences, and they were welcome in their curved forms. Can these unintentional marble details be called material and ornamental details?

---

50 Translated from Turkish to English by the author.

51 Marcel Proust, *In Search of Lost Time*, 7 volumes. (Everyman's Library, 2001[1913–1927]). The titles of the seven volumes are *Swann's Way*, *In the Shadow of Young Girls in Flower*, *The Guermantes Way*, *Sodom and Gomorrah*, *The Prisoner*, *The Fugitive*, and *Time Regained*.

52 Pallasmaa, *The Eyes of the Skin*; Aalto, "Inhimillinen Virhe (The Human Error)": 282; Aalto and Schildt, "Speech at the Centennial Celebration of the Faculty of Architecture."

The Monarga House is not an example of *fragile architecture*, which requires deliberate discontinuities to give a more relaxed aesthetic feeling. "Imperfection is the sign of life ... Irregularities and deficiencies are not only signs of life, but they are also sources of beauty."[53] The Truss Wall House of Ushida and Findlay can be shown as an example for fragile architecture. There is a relationship between the human body and the dynamic flow of the spaces within this building. The building is in a city, but it was designed as a part of nature. The construction techniques used to create the curved surfaces of this building were also unique, but they were *practical*.

"Errors to articulate" are usually applied in art, especially in painting and sculpture. They are not intentional errors, but they are welcome, and they become intentional later. The artist works on those problem parts and articulates them to give an aesthetic character, usually materiality, to that painting or sculpture.[54] Such errors can also be addressed during building processes in architecture. The Monarga House contains two such articulated errors. These are the two rain chains, which solved a big problem in the design of the house. There were two diagonal hip lines on the roof, and when it rained, water flowed into the entrance of the house and the main terrace. By adding these two rain chains, which also have ornamental character, the problem was solved. These two rain chains are *functional as well as ornamental system details*. However, their togetherness with the details of shutters and the leaf plaster may be seen as *overdetailing*.

In Case 1, the OAC described a unique detail that arose due to an error during construction. Although hidden, this detail has a positive tectonic affect, and it resolves a functional problem through an innovative approach. The issue stemmed from a slope problem on an outdoor surface. After the area was covered with ceramic tiles, water began to accumulate and created a potential hazard. Several solutions were attempted, but none worked. Finally, they designed an innovative small groove to drain the water effectively. The OAC expressed satisfaction with this tectonic solution. However, when the RONAC requested a photo of the detail, the OAC explained that it was hidden. This interesting detail combines the characteristics of errors transformed into artistic articulation, innovative solutions, and hidden details. This shows that a hidden detail can also have a positive tectonic affect, evoking certain architectural theories concerning the aesthetics of emptiness and absence.[55] However, these theories primarily pertain to architectural space, not construction details. On the other hand, tectonic theories typically

---

53 Gary Coates wrote that John Ruskin advocated for imperfection, irregularities, and deficiencies in architecture. Coates, *Erik Asmussen.*

54 David Sylvester, *Interviews with Francis Bacon.* (Thames & Hudson, 2016); Neil Leach, *The Anaesthetics of Architecture.* (MIT Press, 1999); Mark Stevens and Annalyn Swan, *Francis Bacon-Revelations.* (Alfred A. Knopf, 2021).

55 Bert Oliver "The Libeskind Jewish Museum in Berlin, the Unpresentable and Experience" *Acta Academica* 53, no. 1, 2021: 23–43.

9 Conclusion

do not consider hidden elements and components as part of tectonic expression. This detail also emphasizes that many unexpected types of details can emerge during the building processes.

The OCCC of Case 2 described a civil engineering detail which was based on an error. This detail was an innovative attempt, and although it was hidden, it caused positive tectonic affects. This happened when the formwork of the reinforced concrete skeletal system was built. The builder made a mistake and, instead of preparing the formwork for an inverted beam, he prepared it for a normal one. The OCCC used this as an opportunity to enlarge the slab of that floor by injecting steel into the concrete and adding a new inverted beam between two columns to carry the heavy wall sitting on that cantilever. As an experienced engineer, he did not want the wall to sit on a cantilever connected to the rest of the structure through injected steel and epoxy. The OCCC described other versions of this type of detail in other buildings he was involved as a civil engineer. This shows that not only architects, but civil engineers are also designers of hidden yet tectonic details. There is also a practical, artistic dimension to civil engineering too.

Errors demonstrate modesty in architecture, and they can be compared with the presence of various levels of architectural *precision*.

- Mathematical perfection in architecture: This approach is against human nature because it does not consider the ergonomics of the human body. Ludwig Wittgenstein designed his sister's house (Hermine's House) according to mathematical proportions and ignored users' functional, ergonomic, and emotional needs. For example, the door handles were positioned in the middle of the doors. Wittgenstein also asked workers to pull down a reinforced concrete slab because it was a few millimeters lower than specified in the architectural project.[56] His sister could not live in this house.
- Precision: Precision in modern architecture is greater than traditional architecture due to the presence of structural engineering applications and industrialization. The search for modular coordination in buildings reflects precision. However, even such precision includes the concept of *tolerance* for the sake of buildability. Mies van der Rohe's architecture can be shown as examples of precision in architecture.
- Order in nature: Order in nature is not mathematically perfect. It is also not precise. Both *hierarchical and rhizomatic orders* exist in nature.[57] Fractal geometry has been developed to understand forms in nature.
- Placement of elements/components approximately: Traditional architecture was not precise. The masons were placing timber posts or stone pieces in

---

56 Stuart Jeffries, A Dwelling for the Gods, *The Guardian*, January 5, 2002, accessed February 22, 2025. https://www.theguardian.com/books/2002/jan/05/arts.highereducation.

57 Deleuze and Guattari, *A Thousand Plateaus*; Deleuze and Guattari, *Anti-Oedipus*.

On a Procedural Theory of Tectonics | **311**

buildings approximately and without taking exact measurements.[58] Many people think that this gives a particular aesthetic character to these buildings.

- Irregularity: The concept of irregularity is used in structural engineering to express some architectural organizations which cause certain earthquake problems. These include non-parallel axes and having some irregularities in the placement of rigid infill wall systems within flexible framed structures. However, architecture is usually irregular, and such problems can be solved by taking some simple precautions.[59]
- Imprecision: Imprecision is different from mistakes because mistakes indicate one point or place in the building, while imprecision is about the weak character of the order of the systems within that building. Fragile architecture suggests imprecision, the presence of deliberate discontinuities, and benign errors in the building. Such imprecision can also be practical. The structural system of the Truss Wall House of Ushida and Findlay, which has curved steel structural elements with shotcrete application, can be shown as an example for imprecision in architecture.

There is no mathematical perfection, irregularity, or imprecision in the Monarga House. The house was typically designed to be precise to a certain extent. For example, the architectural project of this house contained the application of heat insulation on the reinforced concrete frame elements, and this can be realized only with the help of precision. However, this did not happen and the heat insulation layer had to be canceled. There was no proper modular coordination in the Monarga House, and this becomes apparent when the order of the floor ceramics is studied in the completed building. If there was modular coordination, many tiles would not be cut at the corners of rooms. The order of leaf plaster texture on the façades of the Monarga House can be seen as an example of order in nature.

Both precision in the Monarga House, imprecision in fragile architecture, and the approximate placement of elements in the construction of traditional buildings create their own *practicality*. The use of industrialized materials, such as eco-bricks and standard doors, makes precision practical. On the other hand, the imprecision in fragile architecture has similarities to the approximate placement of elements in the construction of traditional buildings. Traditional buildings have natural materials, while fragile architecture uses industrialized materials in a way that doesn't dominate the character of its architecture.

---

58 Lami Eser's explanations of the construction process for timber framed wall structures reveal a lack of precision of such structures. Lami Eser, *Geleneksel ve Gelişmiş Geleneksel Yapı I*. (Istanbul Teknik Üniversitesi Mimarlık Fakültesi, 1977).

59 Yonca Hurol, *The Tectonics of Structural Systems – An Architectural Approach*. (Routledge, 2016).

312 | 9 Conclusion

The issue of practicality can be understood with respect to the following concepts:

- Simplicity
- Complexity
- Being typical
- Practical structure, form, building system, and construction method
- Practicality through technical improvement
- Practicality through innovation.

The typical can be practical because humans usually repeat practical processes. However, one should imagine that there is always a more practical option than the typical. This option makes the minds of architects, controllers and contractors, sub-contractors, builders, foremen, and workers busy, leading to innovative attitudes. Some innovations and improvements are simple, while others are complicated. Innovative attitudes about traditional building techniques are usually simple and can be repeated by almost everybody. Architect Hassan Fathy re-innovated the ancient Egyptian technique of adobe shallow dome building by reading the old Egyptian papyruses. This technique required the use of a simple instrument (the Fathy compass) to make the dome shallow. It was also necessary to use sandbags to keep the adobe bricks in place during construction, eliminating the need for centering. This building technique was so practical and simple that it was easy to understand for ordinary people.[60] However, there are also very complex innovations that cannot be understood easily even by professionals. The use of pneumatic formwork systems for building contemporary reinforced concrete domes can be an example of such complex but impractical innovations. Only specialized engineers can design the structure of these domes and their construction processes. The impracticality of this method is also related to the high cost and unavailability of the pneumatic system for many people. The impractical building processes do not contribute to people improving the building culture within that society because people cannot understand them. Simplicity and complexity can be about form, structural systems, and methods of construction.

Details of all types, as the focal point of procedural tectonics, highlight the contributions of foremen, technicians, and carpenters. Details represent the adoration of fine work, nature, and ethics in building activity.

---

60 Maria Rosaria Gargiulo and Immacolata Bergamasco, The Use of Earth in the Architecture of Hassan Fathy and Fabrizio Carola: Typological and Building Innovations, Building Technology and Static Behaviour, accessed November 13, 2023. https://www.arct.cam.ac.uk/system/files/documents/vol-2-1209-1220-gargiulo.pdf.

A Deeper Inquiry into the Procedural Approach to Theory of Tectonics

Recognition of tectonics as a process may initiate more research in the future. Change management theories can be further developed in accordance with the recognition of inclusive building processes. There is also a requirement for new research on the effects of such procedural changes on related laws, rules, agreements, regulations, and ethics. Additionally, future research should explore the perceptions of builders and foremen regarding the changes they make on construction sites. This research should be grounded in a close relationship between researchers and the construction site. When a change is identified, the individual responsible for the change should be asked questions in order to understand why the change was necessary, the values driving their decision, and how they categorize the outcome. This could contribute to a more detailed theory of the tectonics of details.

Tectonics as a process is more materialistic than tectonics of architecture as a product. The way application project architects, contractors, sub-contractors, builders, foremen, technicians, and carpenters participate in the building process directly affects the materiality of design and the building process. This may improve the tectonics of buildings through hapticity, better details, avoiding instrumentalization, and so on. It can also cause an increase in innovation and improve work ethics because of being more inclusive and collaborative.

Could involving producers such as contractors, sub-contractors, builders, foremen, technicians, and carpenters, who typically lack abstract design tools, in the tectonic design process lead to any unexpected outcomes? How will the meaning of architecture be affected by the materialization of building processes? Will architectural aesthetics, based on abstract principles and symbolism change? Can increasing materiality in building processes help achieve better relationships with nature? Can this approach direct us towards a post-secular architecture?